The Cycle of Emanations
From the Source to the Return
Logan Gray

Summary

Prologue

There is a reality beyond the limits of what your eyes can see. A primordial and eternal force permeates everything that exists, shaping the universe with an invisible and unwavering precision. Since the beginning of time, every star that shines in the sky, every breath you take, every movement in the cosmos is a reverberation of this unchanging, silent, and absolute source. Nothing in the world is random; everything follows an ordered flow that emanates from a singular, immutable, and superior principle. This force, though hidden, guides every aspect of existence, and everything that is emanated carries within it a spark of the power that originated the universe itself.

This book reveals the true nature of this flow. The words that follow do not offer superficial explanations or uncertain theories. What is contained here are immutable truths, transcending common understanding and touching upon a universal knowledge that crosses cultures and ages. The reality described in the following pages is not a distant possibility or a philosophical conjecture, but a faithful depiction of the hidden structure of the cosmos, where every element of creation is part of an eternal cycle of emanation and return to the Source.

The knowledge of emanations, as old as time, has always been present, veiled by layers of mystery, but now it unveils itself before you. Gnosticism, Neoplatonism, Kabbalah—all great esoteric traditions converge at this central point: everything that exists, whether in the material or spiritual plane, is the result of a descent from the primordial unity into the multiplicity of the world. Every form that manifests carries the essence of the One, even if distant from its original purity. Nothing escapes this web of interconnection. Every action, every thought, every existence is linked by an invisible thread to the absolute.

The truth is, you are also part of this grand scheme. You are not a separate or isolated entity in the vast universe. Your very essence is an emanation of this Source, and even though your body and mind are immersed in the density of the material world, within you resides the same force that moves the heavens. The knowledge offered by this book is a key to recognizing this truth and, more than that, to living it. There is no distance between you and the cosmos, between what you perceive as your reality and what lies beyond. Everything is a single current, a single vibration emanating from the deepest center of existence.

By following the ideas presented here, you are not merely reading a text but participating in a much broader cosmic process. Every concept described here reveals the web that connects all phenomena. There is no true separation between the spiritual and the material, between the divine and the human. Emanations permeate everything, and at every level of being, a part of this truth is manifested. What once seemed fragmented now presents itself as absolute unity, where everything has its place and function.

The order of the universe is not random, and every one of its elements, however insignificant they may seem, is part of a greater pattern, of a cosmic dance that pulses continuously. The initial emanation of the One gave rise to all things, and even the densest and most distant layers of this primordial source still carry its light. By understanding this structure, you will begin to see the world differently. What once seemed like chaos or chance will reveal its underlying harmony.

Here, knowledge is not just an abstraction. The truths presented are not conceptual, but living truths that shape every moment of reality. The return to the Source is inevitable, and all things are destined to complete this cycle. There is no true loss, no true separation, only layers of manifestation that, in time, unveil the unity behind the diversity.

You are already immersed in this flow. The path of return has already begun. Emanations are all around you, and the knowledge that this book brings activates within you the ability to perceive them. Here, there is no mystery that cannot be unveiled. As you delve into this knowledge, you will align yourself with this eternal and absolute truth. What lies before you is more than a reading; it is a call to recognize what has always been—the profound and unbreakable unity of all existence.

Chapter 1
Introduction to the Theory of Emanations

In the stillness of the void, there exists a primordial force, an undisturbed origin beyond the confines of thought or language. This is the Source, the Absolute Principle from which all things flow. From it, radiates the grand unfolding of existence—an ancient process known as emanation. The universe, in all its complexity and beauty, is not born through chaos or happenstance but through an ordered cascade, a descent from the highest unity into the multiplicity of forms. Everything that exists, every star, every drop of water, every breath of wind, is but a reflection of that initial pulse of life, an echo of a much greater whole.

To understand emanation, we must first touch upon its most essential structure: the relationship between the Source, that which emanates, and the emanated. The Source, by its very nature, remains unchanging, eternal, and boundless. It stands beyond the limits of human understanding, yet, paradoxically, it is the very root of all comprehension. From this ineffable One, waves of creation spill forth, like ripples expanding outward from a single point in water. Each wave or emanation carries within it a fragment of the Source's essence, but as these waves move further from their origin, they take on distinct characteristics, becoming more defined, more tangible, more distant.

The emanating principle itself remains intricately bound to its offspring, and while the created world seems fragmented, each part retains a trace of the unity from which it came. This interplay of distance and connection is the pulse of the cosmos—a rhythm of emanation and return. From the highest planes of existence to the most grounded material realities, the journey of emanation unfolds in layers, each step a veil draped over the brightness of the original light.

What is crucial in this theory is not simply the creation of things, but their return. All emanated forms, no matter how far they drift, are destined to return to their source, to dissolve once again into the unity from which they were born. This cyclical process suggests that existence is not a random scattering of elements, but a purposeful journey—a dance of expansion and contraction, of outward creation and inward recollection.

At its heart, the Theory of Emanations offers more than just a metaphysical description of the universe. It speaks of a deep, unspoken truth that underlies existence itself: the principle that everything is connected, that no matter how far removed we may feel from the Source, a hidden thread binds us all. To trace the path of emanation is to walk this thread back to the One.

As we delve deeper into the Theory of Emanations, the distinctions between the traditions that have embraced it reveal themselves, each casting the same fundamental truths through a different lens. In Gnosticism, Neoplatonism, and the Kabbalistic traditions, the structure of emanation appears, though each describes it in its own unique way, layering additional mysteries over the original concept.

In Gnosticism, the act of emanation is entwined with the idea of divine knowledge—gnosis. The Source, or the Monad, sends forth a series of divine entities, known as Aeons, each embodying specific aspects of its essence. These Aeons, forming a kind of divine hierarchy, eventually lead to the creation of the material world, a place viewed as distant and flawed compared to the higher spiritual realms. Here, emanation is both a process of creation and estrangement, with the material world reflecting the furthest reaches of the Source's light, now dim and fragmented.

Neoplatonism, most famously articulated by the philosopher Plotinus, portrays a similar model but with a different emphasis. The One, the ultimate principle in Neoplatonism, overflows naturally, giving rise to the Nous, or Divine Intellect, which in turn emanates the World Soul. The process is not a deliberate act, but an inevitable unfolding of the One's perfection. The Nous, contemplating the One, gives birth to multiplicity, while the World Soul introduces time and space, creating a bridge between the eternal and the temporal.

The Kabbalistic tradition, steeped in the mysticism of Jewish esotericism, presents the sefirot as the key players in the emanation process. These ten sefirot are attributes or emanations of God, each representing a different aspect of divine reality. They form the structure through which the infinite essence of God becomes manifest in the world. In this tradition, there is a delicate balance between the descending flow of divine energy and the potential for its return, echoing the cyclical nature of emanation that we have already encountered.

Despite their differences, these traditions share common threads. They all describe reality as existing in layers, each level more removed from the original Source, but still intimately connected to it. The levels of reality, be they called Aeons, Intellectual Principles, or Sefirot, are not separate realms but interconnected stages in the emanation process. What emerges is a universe organized by the principle of cause and effect, where each emanation brings forth another, but none are truly independent of what came before. This interconnectedness suggests that everything in existence, no matter how far removed, carries within it the essence of the Source.

But this process of emanation is not static. It implies movement, a dynamic interplay between the eternal and the temporal, the absolute and the relative. The Source, though seemingly distant, remains ever-present within its emanations. As each level comes into being, it reflects both its distance from the Source and its inherent connection to it. This duality—of separation and unity—drives the entire cosmos. It is a dance of light and shadow, where each form that emerges is, in some sense, a dim reflection of the original light.

The relationship between the Source and its emanations is not one of dominance but of subtle interdependence. Without the emanations, the Source would remain unmanifest, unknowable. And yet, without the Source, the emanations could not exist. This relationship highlights a profound metaphysical truth: creation is not an act of separation, but of continuity. Each layer of reality, though distinct, is not cut off from the others but remains linked in an unbroken chain leading back to the Source.

In these traditions, the universe becomes not a collection of isolated entities but a web of interconnected forces, each playing its part in the cosmic symphony. The further we explore this web, the more we understand the delicate balance of cause and effect, the interplay of unity and multiplicity, and the constant motion between emanation and return. And as we do, we begin to see that the story of creation is not just about the emergence of the many from the One, but about the eventual return of the many back to their origin.

Chapter 2
The Absolute Principle

At the heart of all emanation lies the Absolute Principle—an idea so vast, so profound, that it eludes the full grasp of the human mind. This Principle is the silent wellspring from which everything emerges, yet it itself remains unmoved, untouched by the flow of creation. Describing it is like trying to capture the essence of infinity with a single word. The mystics and philosophers have called it by many names: the One, the Monad, the Source, the Infinite. Yet these are mere symbols pointing to something beyond form, beyond comprehension.

What is essential to grasp about the Absolute Principle is its nature: it is infinite, boundless, and unchangeable. Unlike the world we inhabit, filled with distinctions and movement, the Absolute exists in a state of pure unity. It has no parts, no divisions. It is not bound by time or space. It transcends all dualities—light and dark, good and evil, presence and absence. In this sense, the Absolute is not "something" at all, for to be something implies limitation. It is rather the ground of all things, the hidden source from which all reality springs.

Yet, paradoxically, it is from this very state of undivided unity that all multiplicity arises. The process of emanation begins with the Absolute, and though it remains unchanged, it gives rise to everything. How, we might ask, can the infinite produce the finite? How can the unchangeable bring about change? This is one of the deepest mysteries of the emanation process—a mystery that various traditions have approached with different interpretations, but none have ever fully explained.

The Absolute Principle is often described in negative terms, through what it is not, rather than what it is. This is the apophatic approach, found in mystical traditions from the

Neoplatonists to Christian mystics and beyond. They teach that the Absolute is beyond all attributes. It cannot be defined by qualities such as "good," "powerful," or "wise," for these are attributes that belong to finite beings. The Absolute is beyond being and non-being, beyond light and darkness. It is the ultimate source of all qualities, but it itself possesses no quality that can be comprehended by the mind.

In Neoplatonism, Plotinus speaks of the One as "beyond being," suggesting that it exists in a state of perfection so complete that it transcends even the concept of existence. Similarly, in the Kabbalistic tradition, the Ein Sof—meaning "without end"—describes the Infinite in a state of absolute potentiality, before any distinctions or manifestations have arisen. This idea of an unknowable, ineffable principle can also be found in Eastern philosophies, such as the Tao in Taoism, which is described as the unnameable, the source of all things but beyond all descriptions.

Despite its unknowability, the Absolute Principle is not distant or aloof. It is present in every emanation, every level of creation. Like a mirror, each layer of reality reflects something of the Absolute, though in an increasingly veiled form. The further from the Source, the more the reflection is distorted, but the essence remains. In this sense, the Absolute is both everywhere and nowhere—it transcends all things, yet permeates all things.

One must understand that the Absolute does not create in the way we typically think of creation. It does not form the universe with hands or intention. Instead, emanation is a natural and inevitable outflow from the Absolute's very perfection. Like a flame that cannot help but give off light, the Absolute radiates existence. This is why, in many traditions, the Absolute is seen as impersonal. It does not "decide" to create; it simply is, and from its being, all things flow.

The image often used is that of a fountain or a sun. Just as the sun does not "choose" to shine but simply radiates light by its nature, the Absolute Principle pours forth existence effortlessly. This outpouring is continuous, timeless. It has no beginning or end. The universe, in all its diversity, is simply the unfolding of this eternal act of emanation.

In this way, the Absolute remains unchanged by the process of emanation. It is not diminished by the creation of multiplicity, nor does it become entangled in the forms it produces. The Source remains transcendent, utterly beyond the reach of the emanations that arise from it. And yet, it is immanent within every emanation, present in every particle of creation. It is both the most distant and the most immediate reality.

This dual nature of the Absolute—as both transcendent and immanent—lies at the heart of its mystery. It is the source of all being, yet beyond being. It creates without being changed. It is both utterly unknowable and the most intimate presence in all existence. The Absolute Principle is the key to understanding the entire process of emanation, for it is the origin, the destination, and the thread that connects all things.

Though the Absolute Principle remains unchanged by the emanations it sets into motion, the paradox of its relationship to creation deepens as we try to reconcile its transcendent nature with the immanence found in every particle of existence. How can something so utterly beyond comprehension, beyond time, space, and form, be the very essence that permeates all things? This is the heart of the philosophical and mystical tension surrounding emanation—a tension that has led to profound reflections across different traditions, each offering their own path toward understanding this paradox.

The Absolute, by its very definition, transcends all categories. It is neither "here" nor "there," yet it is everywhere. It does not "become" anything, because becoming implies change, and the Absolute is eternally complete. Nevertheless, everything that exists, everything that moves or grows or fades, does so because of its connection to this Absolute. Each emanation is a whisper of its eternal nature, an echo of its boundless reality, but as these emanations unfold, they move away from the perfect unity of the Source.

This paradox is reflected in the mystical language of many traditions. The Kabbalists, for instance, speak of "tzimtzum," a divine contraction, in which the Infinite withdraws itself to make space for creation. Though the Infinite Light of the Absolute fills all things, this light is progressively concealed as it moves through the layers of emanation. The further the emanations descend, the more veiled the light becomes, until at the material level, the divine presence is almost completely obscured. Yet, even in the most distant realms of existence, the light remains hidden within, waiting to be discovered.

In Neoplatonism, the relationship between the Absolute (the One) and its emanations is framed as a hierarchical unfolding. The first emanation, Nous (Divine Intellect), arises from the contemplation of the One. In this act of contemplation, the Nous mirrors the One, producing a multiplicity of ideas and archetypes—the blueprints of all that will come into existence. Yet, despite this multiplicity, the Nous retains its intimate connection to the One, just as light from a distant star is still part of its source, even if it appears faint to those far away.

The next emanation, the World Soul, represents a further stage of descent. Here, time and space come into play, and the abstract ideas of the Nous begin to take on a kind of life, preparing the way for the material universe. But even at this level, the essence of the Absolute is not lost; it simply becomes more distant, more concealed. The World Soul mediates between the eternal and the temporal, embodying both the unity of the higher realms and the multiplicity of the physical world.

For the mystic, the journey of spiritual ascent is precisely the reverse of this process. Just as the emanations descend from the Absolute into multiplicity, the soul seeks to ascend, peeling back the layers of reality in order to return to the Source. Each layer of reality, from the most material to the most subtle, represents a step on this path of return. The further the soul ascends, the closer it comes to the undivided unity of the Absolute, where all distinctions dissolve, and the knower becomes one with the known.

This ascent is often described as a journey of purification. As the soul moves closer to the Source, it sheds the illusions and limitations imposed by the lower levels of emanation. The material world, with its dualities and distractions, is seen as a veil, and through spiritual practice, the mystic seeks to penetrate this veil, to glimpse the light that shines behind it. The closer one draws to the Absolute, the more these distinctions fade, until, at the highest level, there is only unity—a unity that transcends all thought, all form, all individuality.

But here lies another paradox. Though the Absolute is beyond comprehension, it is also the most intimate aspect of reality. The mystic does not reach the Absolute through distance or separation, but through profound recognition. The Absolute is not somewhere "out there," but within, at the heart of all things. As the Kabbalists suggest, the spark of the Infinite lies hidden within every soul, just as the light of the sun is contained in every ray of light. The path of return is not about escaping the world, but about uncovering this hidden light and recognizing that it was never truly separate from the Source.

This brings us to the delicate relationship between transcendence and immanence. While the Absolute is beyond all things, it is also within all things. It transcends the universe, yet it is closer to us than our own breath. This dual nature can be found in the writings of mystics from every tradition, whether they speak of the divine as an infinite ocean in which we are submerged, or as the innermost center of our being. The mystic's task is not to bring the Absolute closer, for it is already near, but to recognize its presence beneath the layers of existence.

Philosophically, this relationship has been the subject of intense debate. How can the Absolute remain unchanged while still giving rise to the multiplicity of the universe? The answer, though never fully graspable, lies in the nature of emanation itself. The Absolute does not change because emanation is not a movement of creation in time. It is a timeless process, an eternal unfolding. The universe, in all its complexity, exists simultaneously within the Absolute, yet the Absolute remains unbound by the universe. It is as if the Source holds the entire cosmos within itself, but is not confined by it.

Thus, the Absolute Principle stands at the center of all things, yet remains untouched by them. It is both the origin and the destination, the beginning and the end, the eternal now from which all time flows. The journey of emanation, from the highest spiritual realms to the lowest material planes, is but one side of this cosmic rhythm. The other is the return, the process by which all emanations are drawn back to their origin, to the silent, infinite depths of the Absolute. The mystery remains, yet in contemplating it, we come closer to understanding the essence of all existence—the pulse of emanation and return, which moves through every corner of reality, and yet rests, eternally unchanged, in the heart of the Absolute.

Chapter 3
The First Levels of Emanation

From the silence of the Absolute, the first act of distinction emerges. It is a moment without time, a movement without space—a ripple that begins the cascade of existence. This is the first level of emanation, where the Absolute, while remaining untouched and unchangeable, gives rise to the beginning of all that will be. At this level, we encounter the concept of the "Divine Intellect," or what has been called the logos in many traditions. The logos is not a separate entity but the first expression of the Absolute's boundless potential, a manifestation of its infinite perfection.

In this first emanation, the purity of the Absolute is still reflected with great clarity. Here, the unity is not yet broken into multiplicity, but distinctions are beginning to form. The Divine Intellect serves as the mirror through which the Absolute contemplates itself. In this act of self-contemplation, the seeds of all creation are contained. The forms of the universe exist here in their most perfect, undifferentiated state, as archetypes, pure ideas that will later give rise to the multiplicity of the cosmos.

This first emanation is often described in terms of light. Just as light radiates from the sun without diminishing its source, so too does the Divine Intellect emanate from the Absolute, a pure, undiluted reflection of the infinite. In many esoteric traditions, this is the realm where the highest truths are known, not through reasoning or language, but through direct apprehension. It is the source of wisdom, the place where all knowledge is one, unified and unbroken. Everything that will later emerge as distinct—thought, will, form—exists here in its purest potential, as part of the indivisible whole.

In Neoplatonism, this level is called the Nous, the Divine Mind, and it is described as the first reality to emerge from the One. The Nous contains within it the perfect ideas, the eternal forms that shape all things. These forms are not physical, but they are the blueprints of everything that will come to exist in the material world. This level of reality is closest to the Absolute, and though it represents a step away from pure unity, it is still unified in itself, a world of thought before the fragmentation into individual entities.

The Kabbalistic tradition also offers insight into this first emanation. The first of the ten sefirot, Keter, represents the crown or highest point of divine consciousness. Keter is the bridge between the infinite Ein Sof and the rest of creation. It is where the divine will begins to manifest, though still in a state of perfect unity. Within Keter, the potential for all creation exists, but nothing has yet taken form. It is the beginning of divine intention, the first pulse of emanation, still utterly connected to the infinite light of the Source.

What we begin to see here is a pattern of unfolding—each level of emanation arises naturally from the one before it, yet each step represents a further movement away from the pure unity of the Absolute. At the level of the Divine Intellect, this movement is still subtle, almost imperceptible. The distinctions that form here are delicate, like the first glimmers of light before dawn. The multiplicity that will come later is not yet evident. Instead, there is the contemplation of the whole, the presence of all things in their undivided state.

The idea of the logos or Divine Intellect is central to understanding how the cosmos maintains its connection to the Absolute. It is the organizing principle, the bridge between the unknowable infinity of the Source and the world of forms. Through the logos, the Absolute contemplates itself, and this contemplation gives rise to the first distinctions that will eventually become the building blocks of all reality. Yet, even in this process of emanation, the Divine Intellect remains inseparable from the Absolute, like a ray of light that remains one with the sun.

This level of reality also introduces the notion of hierarchy within the process of emanation. Though the Divine Intellect is the first emanation and the most immediate expression of the Absolute, it is still not the Absolute itself. It is the first reflection, the first layer of veiling that begins the process of differentiation. The further we move from this level, the more distinct and defined the emanations become, but here, at the first level, the unity is still evident. It is the source from which all distinctions will flow, but these distinctions have not yet fully taken form.

As we explore the first levels of emanation, it is important to remember that the Absolute does not move or change. The emanations are not a departure from the Source but a natural expression of its boundless nature. Just as a fountain overflows with water or a flame gives off light, the Absolute radiates existence. This emanation is not a creation in the sense of something coming into being from nothing. Rather, it is an unfolding, a gradual revelation of what already exists within the potential of the Absolute.

The Divine Intellect, then, is the first revelation of this potential, the first point where the perfection of the Source begins to take on a recognizable form. It is not yet a world of objects or individual beings, but it is the place where the seeds of all future creation lie. In this sense, the Divine Intellect is both the first emanation and the blueprint for everything that will follow. It contains within it the essence of all that is to come, though at this stage, that essence is still unified, undifferentiated, like light before it passes through a prism.

In this first emanation, we see the beginnings of the cosmic journey—the movement from the infinite to the finite, from unity to multiplicity. It is a journey that will continue as the emanations unfold, each level introducing new distinctions and complexities, yet always retaining some trace of the original unity. The Divine Intellect is the first step in this journey, the first manifestation of the Absolute's eternal perfection.

As the emanation process unfolds, the Divine Intellect, or logos, begins to diversify, bringing with it the seeds of multiplicity. This transition marks a crucial stage in the descent from the Absolute—where unity subtly transforms into the potential for distinct expressions. While the Divine Intellect is still bathed in the light of the Absolute, it now begins to organize and shape the raw, infinite potential into something more defined. The pure, undivided light is about to pass through a prism, splitting into the fundamental structures that will give rise to all reality.

At this stage, distinctions between intellect, will, and action begin to take form. The first emanation contained within it the potential for all things in a unified state, but now this potential must be activated. The Divine Intellect, contemplating the Absolute, moves toward differentiation. It is not yet the world of physical forms, but rather the world of principles and archetypes—the realm of divine ideas that will eventually shape all levels of existence. This is the moment where duality begins to emerge, though it is still in a harmonious balance with the unity from which it sprang.

Within the context of Neoplatonism, this level introduces the interplay between the Nous (Divine Intellect) and the next emanation, the World Soul. The Nous, in contemplating the One, generates a vast array of perfect forms, known as the intelligibles. These are the eternal truths and archetypes that form the blueprint of reality. From the Nous, the World Soul emerges, charged with the task of bringing these archetypes into a dynamic, temporal order. The World Soul represents a further descent, where time and movement come into play, setting the stage for the creation of individual souls and the material world.

In this interplay between intellect and soul, we see the beginning of a duality that will be repeated throughout the process of emanation. The Divine Intellect represents pure thought, while the World Soul introduces the element of will—an active, organizing force that begins to shape the emanations into distinct levels of reality. This duality is not one of opposition but of complementarity. The Divine Intellect remains fixed, eternal, and unchanging, while the World Soul introduces the flow of time, space, and differentiation, allowing the eternal forms to manifest within the finite world.

The Kabbalistic tradition offers another perspective on this stage of emanation. After Keter, the first and most unified of the sefirot, comes Chokhmah (Wisdom) and Binah (Understanding). These two sefirot embody the first major division within the divine mind. Chokhmah represents the flash of pure, undifferentiated insight—an idea in its most potent form. Binah, on the other hand, is the principle of understanding, where this insight is shaped and structured into a more comprehensible form. Together, Chokhmah and Binah mirror the interplay between the Divine Intellect and the World Soul, where the first spark of wisdom is molded into a plan, ready to be unfolded in the subsequent stages of creation.

This duality, whether framed as Intellect and Soul, or Wisdom and Understanding, reflects a deeper truth about the nature of emanation: creation is not a single act, but a series of unfolding stages, where unity gradually becomes multiplicity. Each level builds upon the one before it, retaining some essence of the previous stage but becoming increasingly distinct. The further the emanations descend, the more differentiated and complex they become, yet they remain connected to their source through an invisible thread of light.

As we move into the realm of the World Soul, we begin to encounter the forces of division and multiplicity. The divine archetypes that existed in perfect unity within the Nous are now prepared to take on individual forms. This is where the divine energy begins to split into pairs, opposites, and complements. The masculine and feminine principles, the active and passive forces, the duality of light and shadow—these pairs represent the first significant divisions within the emanation process. Yet, even here, this division is not a break from unity but rather an expression of the underlying harmony that continues to flow from the Source.

In Kabbalah, this division is reflected in the balance between the right and left pillars of the sefirot—the masculine and feminine aspects of divine energy. Chokhmah, associated with the right pillar, is seen as the active, creative force, while Binah, on the left, is the receptive, shaping force. Together, they generate the flow of creation, where the raw energy of insight is tempered and formed into something tangible. This dynamic interplay between opposites is necessary for the process of creation to unfold, as the emanations continue to multiply and diversify.

At this point in the descent of emanation, we also begin to see the emergence of spiritual hierarchies. The higher levels, such as the Divine Intellect and the World Soul, remain closest to the Source and are therefore more unified, while the lower levels start to display a greater degree of separation and complexity. These spiritual hierarchies are often described as chains of being, where each level is dependent on the one above it, but with each step downward, the unity of the Source becomes more obscured.

These hierarchies are not merely abstract concepts but are often personified in many traditions as angelic or spiritual beings. In Neoplatonism, for example, the daimons—intermediate spirits—exist between the world of the divine intellect and the material world. In Kabbalah, the angelic hosts correspond to different sefirot, acting as intermediaries between the divine will and the physical universe. These spiritual beings are the embodiments of the different levels of emanation, serving as channels through which the energy of the Source flows into the lower worlds.

In understanding the first levels of emanation, we come to see that the process is not one of simple division or fragmentation. Rather, it is a dynamic and continuous unfolding, where each new level of reality reflects the unity of the Source while also introducing new distinctions. The Divine Intellect remains the guiding principle, the bridge between the Absolute and the multiplicity that will emerge in the lower worlds. And as these first distinctions take shape, we begin to glimpse the vast hierarchy of being that connects the most subtle spiritual realities to the densest material forms.

At this stage, the emanations are still closely tied to the Source, their unity still evident in the harmonious interplay of intellect and will, of thought and action. But with each step further down the chain, the distinctions will grow sharper, the light more obscured, until the material world—the most distant emanation—comes into view. Even then, the thread of connection remains, a reminder that no matter how far the emanations travel from their origin, they are never truly separate from the Absolute.

Chapter 4
The Multiplicity of Emanations

As the first emanations spread from the Divine Intellect, unity begins to unravel into multiplicity. This is not a chaotic process but a structured, deliberate unfolding, where the simplicity of the original essence gives way to the complexity that will eventually form the universe in all its diversity. The movement from unity to multiplicity is the essential rhythm of emanation, where the single source becomes a myriad of distinct forms, each a reflection of the original light but now separated by layers of differentiation.

In this phase, the intermediate levels of reality begin to take shape. These are not yet the material forms we recognize in the physical world, but spiritual entities—beings of light and consciousness—who exist in realms closer to the Divine Intellect, yet still distinct from it. These levels are often conceptualized as hierarchies, where each stage of emanation exists in relation to the others, with higher levels being more subtle, more unified, and lower levels becoming increasingly differentiated and complex.

Many traditions describe these intermediate levels as a cascade of spiritual hierarchies. In Gnosticism, for example, the Aeons are divine beings who exist in a realm of pure spirit, far removed from the material world. These Aeons are expressions of the Divine Mind, and their emanations form the structure of reality, descending step by step toward the physical plane. Each Aeon contains within it the essence of the one above it, but also adds something new, creating a rich tapestry of spiritual beings that inhabit the cosmos.

Similarly, in Neoplatonism, the procession from the Nous to the World Soul introduces multiple layers of reality. The World

Soul, as the mediator between the eternal and the temporal, is tasked with governing the realm of time and space, but it is supported by various levels of spiritual entities—often referred to as daimons—who oversee the movement of souls and the forces of nature. These beings act as channels for the Divine Intellect, transmitting its order and purpose into the lower realms.

The Kabbalistic tradition offers a detailed view of this multiplicity through its structure of the sefirot. The sefirot are not individual entities but aspects of divine energy, which unfold in a sequence from Keter (the crown) down to Malkhut (the kingdom), where the divine light finally touches the material world. Each sefirah represents a distinct mode of divine expression, such as wisdom, beauty, or justice, and the interplay between them generates the complexity of creation. The further one moves from the higher sefirot, the more differentiated and material the emanations become, leading ultimately to the physical universe.

What emerges at this stage of emanation is the idea that the divine essence is not contained in a single form but is scattered across many forms, each one reflecting a different aspect of the original unity. These intermediate levels are populated by spiritual beings who serve specific functions within the cosmic order. They are guardians of the higher laws, mediators between the Divine Intellect and the material world, and their existence bridges the gap between pure spirit and physical reality.

Yet, as we approach the lower levels of these hierarchies, the light of the Source becomes more and more veiled. The further the emanations move from the Absolute, the more they take on the qualities of separation and differentiation. What was once a single, unified light now becomes fragmented into many rays, each one reflecting a different aspect of the divine but no longer in perfect unity with the others. This fragmentation is necessary for the creation of the material world, where multiplicity and individuality reign.

At the same time, this movement away from unity introduces the potential for imperfection. The further the emanations descend, the more they encounter resistance, limitation, and the possibility of distortion. The energy that was once pure and unbounded now interacts with denser realms of existence, where it is shaped and influenced by the forces of time, space, and matter. These lower emanations are still connected to the Source, but their connection is more tenuous, their light more diffused.

The concept of multiplicity also extends to the structures of the cosmos itself. The universe is not a single plane of existence but a vast, multi-layered reality, where different levels of being coexist, each with its own laws and characteristics. These levels are often described as planes or worlds, and they range from the highest, most subtle spiritual realms to the lowest, densest material ones. The emanations flow through all these worlds, connecting them in a grand, cosmic hierarchy.

In many traditions, these worlds are not merely abstract concepts but are inhabited by spiritual beings who guide and govern the processes of creation. Angels, archangels, and other celestial beings act as intermediaries between the higher and lower realms, ensuring that the flow of divine energy remains intact. These beings are part of the multiplicity that emerges from the unity of the Source, and their roles are essential to maintaining the balance between the spiritual and material dimensions.

This multiplicity, though it appears as a descent from the purity of the Source, is not a loss of connection. Each level of emanation, each spiritual being, each world, still carries within it the essence of the original unity. The further we move down the chain of emanation, the more this essence is obscured, but it is never completely lost. The task of the seeker, the mystic, is to trace this multiplicity back to its source, to see through the layers of differentiation and recognize the underlying unity that binds all things together.

As the multiplicity of emanations prepares the way for the material world, we begin to see the process of creation as a dance between unity and diversity. The Divine Intellect remains the guiding force, ensuring that even in the midst of multiplicity, there is order, harmony, and a return path to the Source. This multiplicity is not a fragmentation in the negative sense but an expression of the infinite potential contained within the Absolute. Each emanation, no matter how distinct, is a part of this cosmic unfolding, a reflection of the divine essence scattered across the vast expanse of creation.

As we descend further into the multiplicity of emanations, we witness how the divine energy begins to shape itself into increasingly distinct and recognizable forms. The transition from pure spiritual light to denser levels of existence introduces complexity, as the energy of the Source becomes more structured, compartmentalized, and defined. These emanations, though still connected to the Absolute, begin to inhabit different planes of reality, each governed by its own laws and hierarchies. At this stage, the spiritual and material worlds are prepared for their emergence, intricately woven together by the descending flow of divine energy.

One of the key aspects of this multiplicity is the gradual development of form. In the higher realms, forms are fluid and subtle, more like archetypal ideas than the fixed, material shapes we recognize in our physical world. These forms—sometimes referred to as divine archetypes or intelligible forms—are the templates from which all things in existence are modeled. They are the first expressions of distinctness within the emanation process, the early signs of separation that will, in time, give rise to the material cosmos.

In Neoplatonism, these forms exist within the Nous, or Divine Intellect, as eternal and unchanging patterns of all that will be. The Nous, in contemplating the One, produces a multitude of these archetypes, each a perfect reflection of the Absolute's infinite potential. But as the emanation process continues, these forms must pass through the World Soul, where they are translated into dynamic, temporal realities. In the World Soul, these eternal forms are given life and motion, becoming the seeds for the physical world.

The concept of spiritual hierarchies becomes even more pronounced here. The spiritual beings inhabiting the higher realms act as intermediaries, maintaining the flow of divine energy as it passes through the various levels of existence. These beings, which in many traditions are identified as angels, archangels, or celestial entities, serve as guardians and guides, ensuring that the divine will is manifested correctly as the energy descends. They are both expressions of the divine multiplicity and participants in the unfolding of creation.

In the Kabbalistic tradition, the idea of multiplicity is illustrated through the sefirot, particularly in how they are arranged into triads and pillars, each interacting with the others to create balance and harmony. As the energy of the divine light flows down through the sefirot, it becomes increasingly differentiated. The higher sefirot, such as Chokhmah (Wisdom) and Binah (Understanding), still reflect the unity of the Source, but as the energy descends, it moves through sefirot like Netzach (Victory) and Hod (Glory), which govern the forces of nature and emotion. By the time the divine light reaches Malkhut, the lowest sefirah, it has become the foundation of the material world—a world where multiplicity reigns.

It is at this stage that the emanations begin to acquire more defined forms, manifesting in denser planes of existence. The spiritual light that once flowed freely through the higher realms is now enclosed within specific structures. These structures are not physical in the way we understand matter, but they represent the organization of spiritual energies into distinct entities. These entities govern the intermediate realms, each carrying out specific functions in the maintenance of cosmic order. They are the administrators of creation, ensuring that the divine energy remains aligned with the original intent of the Absolute.

This process of differentiation also brings with it the notion of hierarchy. As the divine light moves further from its source, it begins to stratify into levels, each one more complex and specific than the last. The higher levels remain close to the Source and are more unified, while the lower levels become increasingly divided and detailed. This hierarchy is not a rigid structure but a fluid, dynamic network where energy flows between the various levels, maintaining the interconnectedness of all things.

In many esoteric traditions, these hierarchical levels correspond to planes of reality, often described as worlds or dimensions. The higher planes are inhabited by beings of pure spirit, while the lower planes become the realms of matter and form. Each plane has its own characteristics and laws, but they are all interconnected, with the energy of the Source flowing through them like a river through a series of cascading waterfalls. The further the energy travels, the more it changes, adapting to the characteristics of each plane, but it always retains a trace of its original purity.

The idea of spiritual hierarchies and planes of existence can also be found in the cosmologies of various mystical systems. In Gnosticism, for instance, the Pleroma is the realm of divine fullness, where the Aeons exist in perfect harmony. Below the Pleroma is the Kenoma, the realm of deficiency, where the material world resides. The emanations must pass through these realms, each step down the hierarchy representing a further separation from the divine source, but also a necessary step in the process of creation.

As the multiplicity of emanations reaches the denser planes of reality, it prepares the way for the material world. The forms that were once fluid and archetypal become more fixed, more tangible. The divine energy, now heavily veiled, is ready to manifest in physical matter. This descent is not a fall or a degradation, but a natural part of the cosmic process—a movement from unity to multiplicity that allows the universe to exist in its current form. The material world, with all its diversity, is the final expression of this multiplicity, the culmination of the descending emanations.

However, as these emanations descend into matter, the connection to the Source becomes more obscured. The divine light is now hidden beneath layers of form and structure, and the beings who inhabit the lower planes may no longer recognize their origin. This veiling introduces the possibility of confusion, fragmentation, and imperfection, as the clarity of the Source is lost in the complexity of the material world. The challenge of existence in these lower planes is to remember the connection to the higher realms, to see through the multiplicity and rediscover the unity that lies at the heart of all things.

In this phase of the emanation process, we see the full unfolding of divine potential. The simplicity of the Absolute has given rise to a universe of incredible diversity, where every aspect of creation is a unique expression of the original light. The multiplicity of emanations, though seemingly distant from the Source, is not separate from it. Each emanation carries within it the essence of the divine, and through the multiplicity of forms, the unity of the Absolute is reflected back in countless ways.

Chapter 5
The Role of Light in Emanations

As we turn our gaze to the symbolic and metaphysical role of light within the Theory of Emanations, we enter a realm where light is more than a physical phenomenon—it becomes the very essence of the divine flow that sustains all creation. Light, in its purest form, represents the first visible expression of the Absolute, the bridge between the ineffable Source and the manifest world. It is through light that emanation flows, illuminating each level of existence, bringing into being the forms and realities that populate the cosmos.

In many esoteric traditions, light is seen as the primary medium through which the divine energy moves. It is the vehicle of creation, the carrier of the divine will, and the force that connects the highest spiritual realms to the lowest material planes. From the earliest stages of emanation, light acts as the first intermediary between the Absolute and the emanations, symbolizing the descent of unity into multiplicity.

In Gnosticism, light is often described as the pure emanation of the Monad—the singular, ineffable Source. The spiritual beings, or Aeons, who inhabit the higher realms are depicted as beings of light, each one a spark of the divine. The material world, in this framework, is seen as the realm where this light has become most fragmented, almost entirely obscured by darkness. Yet, even in the densest corners of the material world, the light of the Source still flickers, hidden beneath the veil of form and matter.

Similarly, in Neoplatonism, light is used as a metaphor for the descent of the Nous (Divine Intellect) from the One. The Nous reflects the light of the One, and through this reflection, the eternal archetypes come into existence. The World Soul, as the next step in the emanation process, further refracts this light, organizing it into the dynamic forces that govern the material universe. In this way, the light of the Absolute passes through each level of reality, becoming more and more diffused as it descends.

In Kabbalah, light plays an even more central role in the process of creation. The divine light, or Ohr, is the essence of the Infinite (Ein Sof) as it moves through the sefirot, shaping the universe. The sefirot themselves are seen as vessels for this light, and each one represents a specific aspect of the divine energy as it descends through the spiritual worlds. At the highest level, the light is pure, undifferentiated, and boundless. As it flows downward through the sefirot, it becomes more structured and defined, until it finally reaches the material world, where it is enclosed within the forms of physical reality.

This light is not simply an abstract symbol; it represents the living, breathing force that animates all things. Every level of existence, every being and object, is a vessel for this light. In the higher realms, the light is more visible, more intense, but as it moves into the lower worlds, it becomes veiled, hidden beneath layers of materiality. Yet, even in its most concealed state, the light remains the essence of all creation. It is the source of life, consciousness, and being.

One of the key aspects of light in the Theory of Emanations is its ability to both reveal and conceal. Light is the means by which the divine is made manifest in the world, but as it descends, it becomes more obscure, more difficult to perceive. This dynamic interplay between light and darkness, revelation and concealment, is central to understanding the process of emanation. The further an emanation moves from the Source, the more the light is hidden, until, at the lowest levels of existence, it seems almost entirely lost.

This theme of concealed light is especially prominent in Kabbalistic thought. The concept of tzimtzum, or divine contraction, describes how the Infinite Light withdrew itself to create the space for the universe to exist. This withdrawal was not a literal absence but a veiling of the divine light, allowing the material world to come into being. The light that remains in the universe is fragmented, scattered, and often hidden, but it is still there, waiting to be rediscovered and reunited with its source.

In many traditions, the journey of the soul is depicted as a journey of light. The soul is seen as a spark of divine light, a fragment of the Absolute that has descended into the material world. Its task is to remember its origin, to recognize the light within itself and within the world, and to return to the Source. This journey is often described as a process of illumination, where the soul moves from darkness to light, from ignorance to wisdom, as it ascends back through the levels of emanation.

Light, therefore, is not only the force through which emanations occur but also the path by which the emanations return to their origin. It is the thread that connects all things, from the highest spiritual realms to the lowest material forms. In this way, light serves as both the vehicle of creation and the promise of return—a reminder that, no matter how far we may seem from the Source, we are never truly separate from it.

The role of light in the emanation process also speaks to the idea of transformation. As the divine light descends, it transforms, taking on new forms and new meanings at each level of reality. At the highest levels, the light is pure potential, infinite and unbound. As it moves downward, it becomes more defined, more particular, shaping itself into the distinct realities that we experience. This process of transformation is essential to the emanation process, as it allows the infinite to become finite, the abstract to become concrete, and the unity of the Source to express itself in the multiplicity of the world.

At the same time, the light retains its connection to the Source. Even in its most concealed form, it carries within it the essence of the Absolute. This means that, in every part of creation, there is a spark of the divine light, a trace of the original emanation that connects it to the Source. The task of spiritual practice, in many traditions, is to uncover this light, to recognize its presence in all things, and to bring it back into harmony with the unity from which it came.

In this way, light becomes not just a metaphor for creation but the very fabric of the universe. It is the medium through which the divine will is made manifest, the force that sustains all existence, and the path that leads back to the Source. Through light, the process of emanation unfolds, and through light, the return to the One is made possible.

As the journey of emanation continues, the divine light that once flowed freely from the Absolute becomes increasingly complex, altered, and concealed as it descends through the different layers of reality. In this phase, we delve deeper into how light is not merely a static force but one that transforms as it travels, becoming dimmer, fragmented, and more intricate the further it moves from the Source. This descent of light, though marked by attenuation, is crucial to the formation of the physical and spiritual worlds, where the balance between light and darkness begins to shape the cosmos.

One of the most important concepts in understanding the role of light at these lower levels is the idea of concealment. In many mystical traditions, light is understood to conceal itself progressively as it emanates further from the Absolute. This concealment is not a negation of light but a necessary step for creation to occur. If the pure, undiluted light of the Source were to remain fully manifest in the lower realms, the distinction between the divine and the created would dissolve, making the multiplicity of the universe impossible.

In Kabbalistic thought, this concealment is represented through the concept of tzimtzum, the divine contraction. The Infinite Light (Ohr Ein Sof) needed to withdraw and limit itself to allow space for creation. This limitation of light created the potential for lower worlds to emerge, worlds where the divine presence is not overwhelming but rather hidden. In this hiddenness, the physical universe and the spiritual planes below the Absolute could form, allowing beings within these realms the opportunity to experience free will, choice, and spiritual growth.

As the light descends through the levels of emanation, it splits and refracts, becoming more differentiated. The light that was once a pure and singular force in the higher realms begins to take on different qualities and intensities. This process is akin to how a beam of light passing through a prism separates into different colors. Each "color" or aspect of light corresponds to a different realm or dimension of existence, where divine energy manifests in distinct forms. These forms, while still carrying the essence of the original light, reflect it in increasingly specific and localized ways.

At each stage, the light becomes more fragmented, giving rise to a greater diversity of forms and entities. In the higher realms, the light may be perceived as purely spiritual, inhabiting beings of light and formless energy. As it continues to descend, however, the light encounters resistance from the lower planes, where the density of matter and the limitations of time and space obscure its brilliance. In these realms, the light is often cloaked in shadow, becoming less recognizable as divine, yet it remains the vital force that sustains all things.

This attenuation of light can also be understood as a process of spiritual distance. In the higher planes, beings are more aware of their connection to the Source because the light is closer to its original intensity. In the lower planes, however, this awareness dims as the light becomes more hidden beneath layers of materiality. The beings in these lower realms may experience separation, forgetfulness, or even the illusion that they are entirely disconnected from the divine. This is one of the central challenges of existence in the material world: to rediscover the hidden light within and to remember the connection to the Source that has never truly been lost.

The theme of concealment and attenuation also gives rise to the concept of "light's concealment" in the spiritual journey. Mystical traditions often describe the path of enlightenment as a process of uncovering this hidden light, lifting the veils that obscure the divine presence within oneself and the world. The concealment of light is not permanent; it is part of the cosmic drama that invites souls to seek out the light, to purify themselves, and to return to a clearer state of being where the divine light is more fully revealed.

In this context, light becomes both a symbol and a reality of spiritual ascent. As souls move upward through the levels of emanation, they encounter progressively purer forms of light, awakening to deeper layers of truth and divine presence. The journey of spiritual awakening is often described as a return to the light, an ascent back through the layers of reality that had once hidden the soul's original connection to the Source. In many mystical traditions, this process involves practices such as meditation, prayer, and contemplation, which help align the soul with the higher planes where the light is more readily accessible.

Another crucial aspect of light in this phase of emanation is its role in creation. In many cosmogonies, light is the active force that shapes the universe, bringing order out of chaos and giving form to the formless. The light of the divine intellect is what allows ideas and archetypes to take shape, and it is through this shaping that the physical universe eventually comes into being. Light, in this sense, is the organizing principle of the cosmos, the force that brings structure and coherence to the multiplicity of creation.

In Gnostic and Kabbalistic traditions alike, light is seen as a creative force that gives birth to both spiritual and material realities. In the higher realms, light remains fluid and dynamic, constantly creating new forms and realities. In the lower realms, where light becomes more rigid and defined, it shapes the material world, bringing about the physical forms we recognize in nature and in the cosmos. Even though the light may be more limited in these lower realms, its creative potential remains active, constantly generating and regenerating the universe.

This creative light also introduces the possibility of transformation. Just as the light can descend and become concealed, it can also be reclaimed and restored. The process of spiritual alchemy, found in many mystical traditions, speaks to this potential for transformation. Through spiritual practice, individuals can purify their own inner light, aligning themselves more closely with the divine and reversing the process of concealment. This inner work is mirrored in the outer world, where the goal is not merely to escape the material plane but to transform it, to uncover the divine light that lies hidden within the fabric of creation.

In this way, light serves as both the creative force of emanation and the vehicle for return. It is the means by which the universe was brought into being, but it is also the path that leads back to the Source. Even in its most concealed forms, the light still carries the potential for revelation, for guiding souls back toward their divine origin. The journey of existence, then, is one of rediscovering this light, of remembering the unity that underlies all multiplicity, and of following the light back to the Source from which it emanated.

Thus, the role of light in the Theory of Emanations is not static; it is a dynamic force that moves through all levels of reality. It flows from the highest planes of existence, where it shines with undiminished brilliance, down to the lower realms, where it becomes hidden and fragmented. Yet, even in its most obscure form, light remains the vital thread that ties all things together, the force that connects the created world to the divine and offers the path of return to those who seek it. The divine light, though veiled, continues to pulse through the fabric of existence, reminding all beings of their origin and their potential to return to the One.

Chapter 6
Emanations and the Creation of the Universe

As the divine light continues its descent through the layers of emanation, the transition from pure spirit to dense matter unfolds. At this pivotal stage, we witness the formation of the universe, where the energies that have passed through the higher realms condense into structures that will eventually form the physical world. This process marks the beginning of a profound transformation—from the abstract to the concrete, from the realm of pure potential to the manifestation of tangible reality.

In the early phases of this descent, the light of the Absolute has already undergone significant attenuation. As it filters through the spiritual hierarchies and becomes refracted by the various levels of emanation, it takes on distinct qualities that prepare it for the creation of the material world. Yet, even in this process of condensation, the light retains its origin in the Absolute, carrying within it the essence of the Source. The physical universe, despite its density and apparent separation from the divine, is still rooted in this original emanation.

The creation of the universe is not a singular event but a gradual process. It is a continuation of the cosmic rhythm of emanation, where each level of reality unfolds from the one before it. In Gnosticism, this process is often framed in terms of a fall—a descent from the Pleroma, the realm of divine fullness, into the Kenoma, the void or deficiency that characterizes the material world. Here, the creation of the physical universe is seen as the furthest point in the emanation process, where the divine light is most concealed and fragmented.

In Neoplatonism, the creation of the material universe is the final act of the World Soul, which organizes the eternal archetypes of the Nous into temporal and spatial forms. The

World Soul, acting as the intermediary between the intellectual and material realms, sets into motion the forces that will give birth to the cosmos. These forces are the manifestations of the divine ideas in space and time, creating the natural world as we know it. The physical universe, therefore, is not a random or chaotic event, but a reflection of the order that exists in the higher, intelligible realms.

Kabbalah offers a similar view, where the creation of the universe is seen as the culmination of the flow of divine light through the sefirot. The divine energy that descends from the highest sefirot eventually reaches Malkhut, the lowest sefirah, where it enters the material world. This process is not one of degradation but of manifestation. The divine light, having passed through the various filters of the sefirot, becomes materialized as the physical universe. In this sense, the material world is both the furthest removed from the Source and the final expression of the divine will.

What emerges from this understanding is a view of creation where the physical universe is a necessary part of the emanation process. The cosmos, with all its diversity and complexity, is not separate from the divine but is an extension of it. The forms that populate the universe—stars, planets, living beings—are all reflections of the divine archetypes that exist in the higher realms. The universe is a mirror of the divine mind, where the ideas of the Nous, or the energies of the sefirot, are given form and substance.

However, the descent of light into matter brings with it certain challenges. As the divine energy becomes more condensed, it encounters the resistance of the material plane. This resistance is part of the nature of matter, which is dense and slow compared to the fluidity of the spiritual realms. The light that once flowed freely in the higher planes now struggles to express itself fully, leading to the fragmentation and obscuration of the divine presence in the physical world. This fragmentation is not just a physical phenomenon but a metaphysical one as well. The beings that inhabit the material world—humans, animals, plants—are all expressions of the divine light, but in a state where that light is often difficult to perceive.

In many traditions, this obscuration is viewed as part of the cosmic drama of existence. The material world is not seen as inherently flawed but as a place where the divine light must work through the limitations of matter. The light is still present in every aspect of creation, but it is veiled, hidden behind the forms and structures of the physical universe. The task of existence, particularly for sentient beings, is to pierce these veils and rediscover the light within. This rediscovery is often framed as a spiritual journey, where the soul must learn to see beyond the appearances of the material world and recognize the divine light that sustains all things.

This process of veiling and fragmentation is mirrored in the cosmogonies of many esoteric traditions. In Gnosticism, the material world is often portrayed as a realm of darkness, where the divine spark is trapped within the prison of matter. The Gnostic myths tell of the soul's descent into this world, where it becomes ensnared by ignorance and illusion. The creation of the universe, in this sense, is both a divine act and a fall, a necessary stage in the soul's journey toward self-realization and return to the Source.

In contrast, Neoplatonism and Kabbalah offer a more balanced view. While the material world is the furthest removed from the divine light, it is not inherently evil or flawed. Instead, it is a place where the divine energy is simply more hidden, more difficult to perceive. The physical universe is still a reflection of the higher realms, but it is a reflection that has been distorted by the limitations of space, time, and matter. The task of spiritual beings, therefore, is to work within these limitations, to transform matter by infusing it with light, and to return it to its original unity with the Source.

This brings us to the role of humanity within the cosmic order. In many traditions, humans are seen as the bridge between the spiritual and material realms. The human soul, as a direct emanation of the Source, carries within it the divine light, even as it inhabits a physical body made of matter. Humanity's unique position in the cosmos allows it to play a crucial role in the process of creation and return. Through consciousness, intention, and action, humans can either perpetuate the fragmentation of light or work toward its unification.

In Kabbalah, this is expressed through the concept of tikkun olam, the repair of the world. Humans, as vessels of divine light, have the ability to restore the broken fragments of light that have become scattered throughout the universe. Through spiritual practices, ethical living, and acts of kindness, humans can gather these fragments and return them to the Source, thus participating in the ongoing process of creation. The universe, in this sense, is not a finished product but a living, dynamic process in which humanity plays a central role.

Thus, the creation of the universe through emanation is not a singular, closed event but an ongoing process. The divine light continues to flow through all levels of existence, shaping and reshaping the cosmos. The material world, though distant from the Source, is still a vital part of this process. It is a place where the divine light, though veiled, can still be found and where the work of spiritual transformation takes place. In this way, the universe is both the final product of emanation and the field in which the return to the Source can begin.

As we move deeper into the discussion of how emanations form the material universe, we begin to explore the mechanisms by which divine light condenses into matter. This process, though mysterious and often cloaked in metaphor, is a crucial part of the unfolding of the cosmos. The transition from pure spiritual energy into the dense, physical world is one of the most significant transformations within the Theory of Emanations, where the infinite essence of the Source becomes the finite and tangible world we perceive with our senses.

In many mystical traditions, the universe is understood as a layered reality, where the higher realms of pure spirit gradually give rise to the lower, denser realms of matter. Each layer of reality is connected by the same divine energy, but this energy takes on different forms depending on its proximity to the Source. The closer the emanation is to the Absolute, the more subtle and unified it remains. As it descends, however, it becomes increasingly differentiated and fragmented, eventually taking on the characteristics of the material world.

One of the key concepts in this process is the idea of "condensation" or "crystallization" of divine energy. In Kabbalah, the descent of divine light through the sefirot is seen as a progressive thickening of energy, where the light becomes more and more structured until it eventually solidifies into matter. This process is not a fall from grace, but rather a necessary unfolding of the divine plan, where the infinite becomes finite and the formless takes on form. The physical world, in this view, is the endpoint of this process, where the divine light is most concealed but still present.

In Neoplatonism, the material world emerges from the World Soul, which takes the eternal ideas of the Nous and organizes them into the temporal and spatial realities of the cosmos. The World Soul acts as the intermediary between the intellectual and material realms, shaping the divine archetypes into the forms that will populate the physical universe. These forms are not random or chaotic, but follow the order and harmony of the higher realms. Even in the material world, there is a reflection of the divine mind, though it is often obscured by the density of matter.

This idea of condensation is also mirrored in modern concepts of energy and matter. In contemporary physics, we understand that matter is essentially condensed energy, with particles and forces interacting to create the solid objects we perceive. In many ways, this scientific understanding parallels the metaphysical idea that the material world is a densification of spiritual light. The divine energy, as it moves through the layers of emanation, becomes more solid, eventually forming the physical world, just as energy condenses into matter in the scientific framework.

The process of emanation leading to creation is not only about the formation of physical matter but also about the structuring of reality itself. Time, space, and the laws of nature are all products of this emanation. In the higher realms, where the light of the Absolute is more apparent, time and space are either nonexistent or radically different from how we experience them. The further down the emanations descend, the more rigid and defined these concepts become. By the time we reach the material world, time and space are fixed frameworks within which physical beings operate. Yet even these laws are reflections of higher principles, echoes of the divine order that governs the entire cosmos.

In Gnosticism, the creation of the material world is often described as a result of a further separation from the divine light. The myth of the Demiurge, a lower, imperfect creator being, reflects the idea that the material universe, while still a product of divine emanation, is marked by fragmentation and distortion. The Demiurge, often depicted as ignorant of the higher realms, fashions the physical world without full awareness of its connection to the divine Source. In this view, the material world is a realm of shadows, where the light of the Pleroma is faint and obscured, but not entirely absent.

However, in other traditions, such as Kabbalah and Neoplatonism, the creation of the universe is seen as a harmonious unfolding of divine will. The physical world, though distant from the Source, is still a necessary part of the cosmic order. It is the arena in which the divine light can manifest in its most differentiated form, where the multiplicity of creation reaches its fullest expression. The task of spiritual beings, particularly humans, is to recognize this light within the material world and work to bring it back into alignment with the higher realms.

This recognition is part of the broader cosmic process of return. Just as the divine light descended through the layers of emanation to create the universe, it also seeks to ascend, to return to its original unity. This return is not a negation of the material world but a transformation of it. Through spiritual practice, meditation, and acts of creativity, humans can lift the veil of matter and reveal the light that lies hidden within. This idea is central to the concept of spiritual alchemy, where the base elements of the physical world are transmuted back into their divine essence.

The parallels between ancient metaphysical ideas and modern physics deepen when we consider the relationship between energy and consciousness. In many mystical traditions, consciousness is seen as a key element in the process of creation. The divine mind, or Nous, does not simply generate the forms of the universe passively; it actively contemplates and organizes them. Similarly, in contemporary science, consciousness is increasingly seen as a fundamental aspect of reality, with some theories suggesting that the universe itself may be shaped by conscious observation.

In this light, the creation of the universe through emanation can be seen as an act of divine consciousness, where the universe is both an expression of and a reflection of the divine mind. The forms that emerge in the material world, whether physical objects, living beings, or even thoughts, are all manifestations of this divine consciousness. The universe, therefore, is not a static creation but a living, breathing entity, constantly unfolding and evolving as the divine light moves through it.

This view of the universe as a dynamic process rather than a fixed object aligns with both mystical and scientific perspectives. In mystical traditions, the universe is seen as constantly evolving, with new emanations flowing from the Source and new forms emerging in the material world. Similarly, in modern cosmology, the universe is understood to be in a state of constant expansion, with galaxies, stars, and planets forming and reforming over billions of years. This dynamic process reflects the ongoing flow of divine energy, which continues to shape the cosmos even after the initial act of creation.

The creation of the universe through emanation is, therefore, not a one-time event but an ongoing process. The divine light continues to move through all levels of reality, bringing new forms into existence and transforming old ones. The material world, far from being a dead or static place, is alive with the energy of the Source, constantly in flux as it reflects the ongoing emanation of divine will.

In this way, the material universe becomes both the product of emanation and the field in which the return to the Source takes place. It is a place where the divine light is both concealed and revealed, where the infinite becomes finite but never loses its connection to the Absolute. Through the act of creation, the universe unfolds, and through the process of return, it seeks to reunite with the One from which it came.

Chapter 7
Spiritual Beings in the Emanations

As the divine light descends through the layers of reality, it gives birth not only to worlds but also to the spiritual beings who inhabit them. These beings are not merely products of the emanation process—they are active participants in it. As intermediaries between the higher, more unified realms of pure light and the lower, denser material worlds, these spiritual entities serve a crucial role in maintaining the balance and flow of divine energy throughout creation.

In the higher levels of emanation, closest to the Source, these beings are often described as beings of pure light or consciousness. In Gnostic traditions, they are known as Aeons, divine entities that reside in the Pleroma, the realm of fullness. These Aeons embody aspects of the divine mind and, like mirrors, reflect the pure essence of the Absolute, untainted by the imperfections of the material world. Each Aeon represents a different aspect of divine knowledge, such as Wisdom (Sophia) or Thought (Ennoia), functioning as both guardians of these qualities and conduits for the divine energy that sustains the cosmos.

In Neoplatonism, spiritual beings play a similar role, acting as intermediaries between the Nous and the material world. Plotinus speaks of daimons, spiritual entities that govern the natural forces of the universe and the movements of individual souls. These beings are not entirely separate from the divine mind but are expressions of it, emanating from the Divine Intellect and tasked with guiding the flow of energy from the spiritual realms into the material world. They maintain the order of creation by ensuring that the divine ideas are properly manifested within the physical universe.

In the Kabbalistic tradition, the spiritual beings that inhabit the various levels of emanation are often identified with the angelic hosts that correspond to the sefirot. The sefirot, which are channels of divine energy, are not only abstract principles but also inhabited by angelic beings who represent and embody these qualities. For example, the archangel Metatron is often associated with Keter, the highest of the sefirot, symbolizing the direct connection to the Infinite Light. Similarly, the archangels Gabriel, Michael, and Raphael are linked to other sefirot, each playing a role in transmitting the divine will to the lower realms.

These spiritual beings exist not only to maintain the structure of the cosmos but also to act as intermediaries between the Source and the various layers of creation. They serve as conduits for the divine light, ensuring that the energy of the Absolute flows smoothly through the hierarchy of existence. The closer these beings are to the Source, the purer and more unified they remain. However, as the light descends into the lower realms, the beings that inhabit these levels become more differentiated, reflecting the increasing complexity and fragmentation of the emanations themselves.

One of the key roles these spiritual beings play is to guide and oversee the process of creation. In the higher realms, they assist in the formation of the divine archetypes—the eternal forms that serve as the blueprint for all that exists. These beings, sometimes referred to as "intelligences" or "principles," ensure that the divine ideas are properly organized and transmitted through the various levels of reality. Their existence is not separate from the emanation process but is an integral part of it, as they act as the means by which the divine will is expressed in the universe.

In the lower realms, these beings become more involved with the material world. As guardians and guides, they interact with the natural forces of the cosmos, ensuring that the laws of nature are upheld. They also serve as protectors and guides for individual souls, helping them navigate the complexities of the physical world and the journey of return to the Source. These beings of light, though more distant from the Absolute than the higher Aeons or intelligences, are still deeply connected to the divine flow of energy, acting as bridges between the spiritual and material worlds.

In many traditions, spiritual beings also play a role in the moral and ethical dimensions of creation. They are often seen as the enforcers of cosmic justice, ensuring that the balance of light and darkness, order and chaos, is maintained. In this sense, they serve not only as protectors of the divine order but also as agents of spiritual evolution. They guide souls toward enlightenment, helping them recognize the divine light within themselves and work toward their eventual return to the Source.

However, these spiritual beings are not static or unchanging. As participants in the emanation process, they are constantly evolving, just as the universe itself evolves. Their roles may shift as the cosmic order unfolds, and they may take on new functions as the emanations continue to descend and ascend. Some traditions, particularly in Gnosticism, even speak of certain spiritual beings falling into states of imperfection or error. The myth of Sophia, for example, describes how she strayed from the Pleroma, creating a disruption in the cosmic order that ultimately led to the creation of the material world. This fall introduces the idea that even spiritual beings, despite their closeness to the Source, are capable of error and must also seek redemption and return to unity.

In this way, the spiritual beings in the emanation process mirror the journey of human souls. Both are emanations of the divine light, and both must work to maintain their connection to the Source, even as they operate within the lower, more fragmented realms of existence. Just as humans are tasked with recognizing and restoring the divine light within themselves, these beings are tasked with ensuring that the flow of divine energy remains aligned with the original will of the Absolute.

As intermediaries, spiritual beings also play a vital role in facilitating communication between the higher and lower realms. In many mystical traditions, they serve as messengers, carrying divine wisdom to those who seek it. These beings often reveal themselves to mystics, prophets, and sages, offering guidance and insight into the deeper workings of the universe. Their presence is a reminder that, despite the apparent distance between the material world and the divine, the connection between the two remains intact, and the light of the Source continues to flow through all levels of creation.

Ultimately, the role of spiritual beings in the emanation process is one of guardianship, guidance, and evolution. They are the caretakers of the cosmic order, ensuring that the divine light reaches all corners of the universe, even in its most concealed forms. Through their efforts, the balance between light and darkness, unity and multiplicity, is maintained, allowing the universe to unfold in harmony with the will of the Absolute. Their existence reminds us that the process of emanation is not a solitary journey but a collaborative effort, where beings of light, both seen and unseen, work together to uphold the divine plan.

As we deepen our exploration of the role of spiritual beings within the vast system of emanations, it becomes clear that these entities, while acting as intermediaries between the Source and creation, also serve critical functions in maintaining cosmic balance. Their roles are not merely static; these beings actively shape the flow of divine energy across the realms of existence, contributing to the delicate equilibrium between the spiritual and material worlds. At this level of complexity, their influence extends beyond the abstract and becomes intertwined with the very structure of human experience, the organization of the cosmos, and the soul's journey toward the Source.

One of the key aspects of spiritual beings that emerges in many esoteric traditions is their hierarchical organization. From the highest, most sublime beings closest to the Absolute to the lower spirits that interact more directly with the material world, the hierarchy of these entities mirrors the layers of emanation itself. This hierarchy is not merely one of power or authority but of proximity to the divine light. The closer a spiritual being is to the Source, the purer its essence, and the more profound its connection to the original emanation.

In Kabbalistic thought, these beings are often associated with the sefirot, each angelic entity embodying the qualities and attributes of the divine energy flowing through a specific sefirah. These angels and archangels are not abstract figures—they act as the very expression of the divine forces, ensuring that the will of the Absolute is carried out across the various levels of existence. Archangels like Metatron, associated with the highest sefirot, embody the transmission of divine will in its purest form, while other angelic beings, such as Gabriel and Michael, are responsible for the execution of that will within the realms closer to the material plane.

In this system, spiritual beings also play a vital role in maintaining the coherence of creation. The further one moves from the Source, the greater the possibility for chaos, distortion, and imbalance. These beings work to prevent the fragmentation of the divine light, guiding the energies as they descend through the spiritual worlds and into the material universe. Their roles are not passive—they act as protectors of the cosmic order, ensuring that the flow of energy remains aligned with the original intent of the emanations.

In Neoplatonism, we encounter a similar idea with the daimons or intermediary spirits, which govern the natural order and the movements of the souls. These spiritual beings are responsible for overseeing the interactions between the World Soul and the material universe. Their task is to ensure that the divine ideas, which exist in their most perfect form in the higher realms, are properly translated into the natural and human realms. This guidance is subtle but powerful, often influencing the course of events in ways that maintain the underlying harmony of the cosmos.

In Gnostic traditions, where the material world is often seen as a more flawed or deficient realm, spiritual beings take on an even more urgent role. The Aeons in the Pleroma, the divine fullness, are responsible for maintaining the integrity of the higher realms, while their influence filters down into the material world, which is often perceived as a realm of shadows and imperfection. Gnosticism introduces the idea that certain spiritual beings, such as the Demiurge, may act in opposition to the higher divine will, creating a distorted version of the original emanation. This idea of spiritual beings embodying both light and dark aspects becomes central to the Gnostic understanding of the universe, where the material world is a battleground for the restoration of divine balance.

Beyond their roles in maintaining cosmic order, spiritual beings are also deeply involved in the evolution of individual souls. In many traditions, angels and other higher entities serve as guardians and guides for human souls, helping them navigate the challenges of the material world and reminding them of their divine origin. These beings often intervene in subtle, invisible ways, offering protection and guidance to those who seek the light. In mystical traditions, individuals who embark on spiritual journeys frequently report encounters with spiritual beings, who provide insights and direction that help them transcend the limitations of the material realm.

This connection between spiritual beings and human souls reflects a profound interdependence. Just as spiritual beings help guide souls toward the light, human actions can influence the spiritual realms as well. In Kabbalistic thought, humans are seen as crucial participants in the process of tikkun olam, or the restoration of cosmic balance. Through their actions—both physical and spiritual—humans can help repair the fractured elements of creation, elevating the divine sparks that have become trapped in the material world. Spiritual beings, in turn, assist in this process, guiding humans as they seek to realign themselves with the higher emanations.

These beings are also responsible for mediating between the different dimensions of existence. Acting as bridges between the spiritual and material worlds, they facilitate communication between the realms of pure spirit and those bound by physical laws. In this way, they ensure that the divine energy does not become stagnant or blocked, allowing the continual flow of light from the Source to permeate all levels of creation. Their presence reinforces the idea that the universe is a dynamic, interconnected system, where the movement of energy is constant and reciprocal.

Another critical role of spiritual beings is in the administration of divine justice and the balance of karma. Many traditions hold that angels and other spiritual entities are responsible for maintaining the balance between good and evil, light and darkness. They observe human actions, influencing outcomes in ways that align with the greater cosmic order. This function is particularly evident in the idea of guardian angels or personal guides, who assist individuals in making choices that align with their higher selves and the greater good.

This role is not simply reactive; spiritual beings are actively involved in the unfolding of destiny. In some cases, they are seen as the agents through which divine providence operates, ensuring that events in the material world follow a pattern that aligns with the overarching divine plan. They help to orchestrate circumstances in subtle and often unseen ways, creating opportunities for growth, learning, and ultimately, the return to the Source.

In many mystical experiences, spiritual beings appear to guide the soul through moments of transition—birth, death, and significant spiritual awakenings. These beings act as escorts between the realms, ensuring that souls move smoothly from one state of existence to another. In particular, during death, spiritual beings are often described as guiding souls through the liminal space between the material world and the afterlife, helping them to transition from the confines of the body to the more fluid realms of spirit.

This role of guiding souls extends beyond individual lives into the broader cycles of existence. Spiritual beings are also involved in the process of reincarnation or spiritual rebirth, helping souls to choose their next incarnation based on their karmic needs and their path of spiritual evolution. They ensure that the soul's journey remains aligned with its ultimate goal: to return to the Source and reunite with the divine light from which it originated.

As we consider the full scope of spiritual beings in the Theory of Emanations, their role becomes clear: they are the custodians of both cosmic and individual balance, ensuring that the flow of divine light is maintained and that souls can navigate the complexities of the material world. They act as both guides and guardians, working invisibly to keep the connection between the spiritual and material realms intact. Their existence within the hierarchy of emanation is a reflection of the layered nature of the universe, where light and energy flow continuously from the Absolute to the furthest reaches of creation, and back again.

Ultimately, the presence of spiritual beings serves as a reminder that the universe is alive with unseen forces, constantly working to maintain the harmony of the emanations and to help souls rediscover their connection to the Source. Their influence, though often subtle, is woven into the fabric of existence, ensuring that the divine light continues to permeate all levels of creation, even in the darkest and most distant realms.

Chapter 8
The Emanation of the Human Soul

In the unfolding process of emanation, the human soul holds a unique and profound place. Unlike other beings that exist within the layers of spiritual hierarchies, the human soul is not merely a distant echo of the Source, but a direct emanation of it. It carries within itself the divine spark, a fragment of the Absolute, and, in this sense, is both bound by the laws of the material world and eternally connected to the divine light from which it originated.

The emanation of the human soul begins in the highest realms, where it exists in a state of pure unity with the Source. In this primordial state, the soul is undivided and formless, a part of the infinite ocean of divine consciousness. Yet, as the process of emanation unfolds, the soul descends through the layers of reality, taking on distinct characteristics and identities, much like the light of the Absolute itself, which becomes increasingly fragmented as it moves further from its origin.

In many esoteric traditions, the descent of the soul is described as a journey from the highest spiritual realms into the lower, denser realms of existence. In Gnosticism, for instance, the soul's descent is often portrayed as a fall from grace, where the pure light of the divine becomes entangled in the material world. This fall is not a punishment but a necessary part of the cosmic process, where the soul, by inhabiting a physical body, experiences separation from the divine and begins the journey of remembering and returning to its Source.

Similarly, in Neoplatonism, the soul is seen as an emanation of the Nous, the Divine Intellect. In its highest form, the soul contemplates the eternal truths of the divine mind, but as it descends into the material world, it becomes subject to the limitations of time, space, and matter. This descent introduces the dual nature of the human soul: it is both a part of the eternal and unchanging realm of the divine, and yet it must operate within the ever-shifting and imperfect realm of the physical world. The soul, in this context, is tasked with bridging these two realms, carrying the light of the divine into the world of forms while seeking to return to the purity of its original state.

In Kabbalah, the journey of the soul is intricately tied to the structure of the sefirot. The human soul is said to pass through the various sefirot as it descends from the Infinite (Ein Sof) into the material world. Each sefirah imparts different qualities to the soul, shaping its journey and its role within creation. At its highest level, the soul is connected to Keter, the crown of divine will and wisdom, but as it moves downward through the other sefirot, it becomes increasingly defined and individualized, eventually manifesting in the physical world as a human being. In this way, the soul retains its connection to the divine, even as it takes on the qualities necessary to function within the material world.

The human soul, therefore, is both an emanation and a vessel for the divine light. It is imbued with the potential to reflect the divine will, but it must navigate the complexities of the material world, where the light is often obscured by the illusions of separation and duality. The physical body, while necessary for the soul to exist in the material plane, acts as a veil that hides the soul's true nature. The process of incarnation, where the soul enters a body, is often seen as a state of forgetting—where the soul's original connection to the Source is hidden behind the distractions and limitations of earthly life.

This forgetting is not permanent, however. In many spiritual traditions, the soul's journey through life is seen as an opportunity to rediscover its divine origin. The soul, through experiences of suffering, joy, and growth, begins to remember its true nature. Mystical practices such as meditation, prayer, and contemplation are often viewed as tools that help the soul strip away the layers of illusion and reconnect with the divine light within. In this sense, human life is not a random series of events but a purposeful journey, where each experience serves as a stepping stone toward spiritual awakening.

In Gnostic thought, this journey is often depicted as a return to the Pleroma, the fullness of divine light. The human soul, trapped in the material world, seeks to ascend back through the layers of reality, shedding the illusions of the physical world and returning to its original state of unity with the divine. This ascent is not easy, as the material world is filled with distractions and obstacles that prevent the soul from seeing the light within. Yet, through gnosis—direct knowledge of the divine—the soul can begin to transcend these limitations and awaken to its true nature.

In Neoplatonism, the return of the soul to the Source is described as a process of purification. The soul, having descended into the material world and become entangled in its complexities, must undergo a process of spiritual purification to return to its original state of contemplation and unity with the Divine Intellect. This process involves the soul's ascent through the levels of reality, shedding the attachments and desires that bind it to the material world, and reawakening its ability to perceive the eternal truths of the higher realms. In this way, the soul's journey is one of transformation, where it moves from a state of forgetfulness and fragmentation back to a state of divine unity.

Kabbalistic teachings also emphasize the soul's journey as a process of returning to the divine. The soul, having descended through the sefirot, must now ascend back through these same levels to reunite with the Infinite Light. This ascent is often described in terms of tikkun, or spiritual repair, where the soul works to heal the fragmentation caused by its descent into the material world. Through acts of kindness, spiritual practice, and the pursuit of wisdom, the soul participates in the cosmic process of restoration, helping to reunite the scattered sparks of divine light with the Source.

The dual nature of the human soul—its connection to the divine and its embodiment in the material world—also speaks to the human experience of free will. Unlike other spiritual beings, which are more directly aligned with the flow of divine will, the human soul has the capacity to choose its path. It can either align itself with the higher emanations, working toward its return to the Source, or it can become more deeply entangled in the illusions of the material world, forgetting its divine origin. This choice is one of the central dramas of human existence, where the soul's potential for spiritual awakening is balanced against the temptations and distractions of earthly life.

Yet, even in moments of forgetfulness, the divine light within the soul never truly fades. It remains a constant presence, a spark that can be reawakened through spiritual practice, love, and the pursuit of truth. The soul, in its essence, is always connected to the Source, and no matter how far it may seem to stray, the path of return is always open.

In this way, the human soul is not only an emanation of the divine but also a key participant in the cosmic drama of emanation and return. Through its journey in the material world, the soul has the opportunity to rediscover its divine nature, to awaken to the light within, and to assist in the restoration of the cosmic order. Each soul carries within it the potential for transformation, not only of itself but of the world around it, as it brings the light of the divine into the material plane and works to reunite all of creation with the Source.

As we delve deeper into the nature of the human soul and its connection to the process of emanation, we encounter a more detailed exploration of the soul's structure, its levels, and its path of return to the Source. The human soul, far from being a singular entity, is multi-layered, with each layer corresponding to different aspects of the emanation process. These layers reflect the journey the soul undertakes, from its origin in the highest spiritual realms to its incarnation in the material world and, ultimately, its potential return to the Absolute.

In many esoteric traditions, the human soul is seen as composed of several distinct parts, each linked to a different level of reality. In Kabbalah, for example, the soul is divided into five levels: Nefesh, Ruach, Neshamah, Chayah, and Yechidah. Each of these levels corresponds to different stages of the soul's journey through the emanation process and its connection to the divine. Nefesh, the lowest level, is closely tied to the body and the material world. It is the animating force that sustains physical life but is also the aspect of the soul most deeply entangled in the illusions and desires of the material realm. As the soul ascends through the levels, it moves through Ruach, the seat of emotion and intellect, and Neshamah, the higher consciousness connected to spiritual wisdom and divine understanding. At its highest levels, Chayah and Yechidah, the soul reaches a state of pure connection to the divine, where it merges with the infinite light of the Source.

This multi-layered structure of the soul reflects the complexity of its existence. While the soul operates within the constraints of the physical world, it simultaneously retains a connection to the higher spiritual realms. The higher levels of the soul—those closest to the Source—remain untouched by the material world, even as the lower aspects become entangled in the physical body and its experiences. This dual nature of the soul is the key to understanding the spiritual journey: the soul must rediscover and awaken its higher aspects, which are always connected to the divine, even when they are obscured by the distractions of earthly life.

The journey of the soul through these levels is often described as a process of spiritual refinement and purification. As the soul engages with the material world, it accrues layers of forgetfulness and attachment, which veil its connection to the Source. These layers must be stripped away through spiritual practice, ethical living, and self-awareness, allowing the higher aspects of the soul to emerge and shine through. This process is not only one of individual growth but also one of cosmic significance. In many traditions, the purification of the soul is seen as contributing to the larger process of cosmic repair, where the divine light that has been scattered through the material world is gathered and returned to its origin.

In Gnostic thought, the soul's journey is framed as a struggle to escape the material world and return to the Pleroma, the realm of divine light. The material world, governed by the Demiurge, is seen as a place of ignorance and entrapment, where the soul forgets its true nature. Yet, within the soul lies a spark of the divine, a fragment of the original light that longs to return to the Pleroma. The soul's task is to awaken this spark through gnosis, or direct knowledge of the divine, and to transcend the illusions of the material world. In this view, the soul's journey is not just a personal quest but a cosmic battle between light and darkness, where the soul's liberation contributes to the restoration of the divine order.

Similarly, in Neoplatonism, the soul's journey is described as a return to the Nous, the Divine Intellect, from which it originated. Plotinus speaks of the soul as descending into the material world, becoming entangled in the sensory experiences and desires of earthly existence. Yet, the soul always retains a memory of its origin in the higher realms. Through philosophical contemplation and the practice of virtue, the soul can gradually free itself from the attachments of the material world and ascend back toward the Nous. This process is one of recollection, where the soul remembers its true nature and reawakens its connection to the eternal truths of the divine mind.

The Kabbalistic concept of tikkun, or cosmic repair, is closely linked to this journey of the soul. In this tradition, the soul is seen as a participant in the restoration of the divine order. Through its actions in the material world, the soul has the power to gather the scattered sparks of divine light and return them to the Source. Every act of kindness, every moment of spiritual awakening, contributes to this process of repair, bringing the universe closer to its original state of unity. The soul's journey is, therefore, not only about its own return to the divine but also about helping to heal the fractures in the cosmos caused by the descent of the light into matter.

The idea of reincarnation, found in many mystical traditions, also plays a significant role in the soul's journey. In

Kabbalah, reincarnation, or gilgul, is seen as a means for the soul to complete its mission in the material world. If the soul fails to fulfill its purpose in one lifetime, it may return in another, carrying with it the lessons and unresolved issues from previous incarnations. This cycle of reincarnation is not seen as a punishment but as an opportunity for the soul to continue its work of spiritual refinement and cosmic repair. Each incarnation offers the soul a new chance to align itself with the divine will and to advance on its path of return to the Source.

In some traditions, the soul's journey through multiple incarnations is also linked to the idea of karma, the law of cause and effect. The actions of the soul in each lifetime create karmic imprints that shape its future experiences. Positive actions, aligned with the flow of divine light, help the soul ascend, while negative actions, driven by ignorance and attachment, bind the soul more tightly to the material world. The process of reincarnation, therefore, allows the soul to work through its karma, gradually purifying itself and moving closer to spiritual liberation.

The soul's potential for return to the Source is not a predetermined outcome but depends on the choices it makes throughout its journey. Free will plays a central role in this process. The soul, though shaped by its experiences and karmic imprints, always retains the ability to choose its path. It can either become more deeply enmeshed in the material world, losing sight of its divine origin, or it can seek to awaken and realign itself with the higher emanations. This choice is the essence of the spiritual path, where each soul must decide whether to pursue a life of attachment or a life of spiritual awakening.

As the soul ascends through the levels of reality, it undergoes a transformation in its very essence. The lower aspects of the soul, which are bound by the limitations of time, space, and matter, begin to dissolve, allowing the higher, more refined aspects to emerge. This transformation is often described in terms of enlightenment, where the soul reaches a state of unity with the divine light and transcends the dualities of the material world. In this state, the soul no longer perceives itself as separate from the Source but recognizes its inherent connection to the divine. The soul, having completed its journey of emanation and return, merges with the Absolute, becoming one with the infinite light from which it originated.

In this ultimate state of return, the soul experiences not only personal liberation but also a profound sense of fulfillment in its role within the larger cosmic process. The journey of the soul, from its descent into the material world to its ascent back to the divine, is a microcosm of the larger process of emanation and return that governs the universe. Each soul, through its individual journey, contributes to the overall movement of the cosmos, helping to restore the unity of the divine light and bring all things back to the Source.

Thus, the human soul, though seemingly small in the vastness of the cosmos, plays a crucial role in the divine drama of creation and return. Through its journey, the soul not only seeks its own reunion with the Source but also participates in the restoration of the entire universe. Each step on the soul's path—whether one of descent into the material world or one of ascent toward the divine—is a reflection of the larger cosmic cycle of emanation, fragmentation, and return to unity.

Chapter 9
The Cycle of Return of Emanations

As we move deeper into the exploration of emanation, it becomes clear that the process is not one of simple descent into multiplicity, but a cyclical journey—where the emanations, after flowing out from the Source, naturally seek their return. This return is essential to the metaphysical balance of the cosmos, as it mirrors the ongoing rhythm of creation and dissolution, expansion and contraction. The cycle of return reflects the inherent drive of all things, both spiritual and material, to reunite with the Absolute from which they originated.

The cycle of return begins as soon as the process of emanation is set into motion. Just as light flows outward from the Source, dispersing through the layers of existence, it simultaneously carries within it the impulse to return. In this way, every emanation, whether a spiritual being, a human soul, or a material form, is part of a vast, interconnected system that continuously moves toward reintegration with the original unity. The descent into multiplicity, with its attendant complexities and separations, is only one half of the cosmic process; the return to unity completes it.

In Neoplatonism, the return of emanations to the One is central to the metaphysical framework. Plotinus describes the One as both the source and the ultimate destination of all existence. The descent of emanations into the lower realms, though necessary for the existence of the cosmos, is also marked by an innate longing to return to the purity of the One. This return is not a linear journey but a cyclical one, where each soul or emanation moves through stages of purification, gradually shedding the layers of illusion and fragmentation that came with its descent into the material world. As the soul or emanation ascends, it remembers its original state of unity and seeks to reunite with the divine.

In Gnostic traditions, the return of emanations is often framed as a process of awakening from the illusion of separation. The material world, governed by the Demiurge, is seen as a place where the divine light is hidden, trapped in the darkness of ignorance and illusion. The soul, having descended into this world, must begin the journey of return by seeking gnosis—direct knowledge of the divine. This knowledge allows the soul to break free from the material realm and ascend back through the layers of emanation, returning to the Pleroma, the fullness of divine light. In this context, the return is both an individual and a cosmic event, where the restoration of the soul mirrors the restoration of the entire universe.

The concept of return is also deeply embedded in Kabbalistic thought. The cycle of emanation and return is reflected in the structure of the sefirot, where divine light flows outward from the Infinite (Ein Sof), through the various levels of reality, and then seeks to return to its source. In this system, the descent of light into the material world is not seen as a fall, but as a necessary part of the cosmic process. The light becomes fragmented and hidden within the material world, but it retains the potential to return to its original state of unity. The practice of tikkun, or spiritual repair, is the process by which this return is facilitated. Through the actions of human beings, the divine sparks scattered throughout creation can be gathered and restored to the Source.

In this cycle, human souls play a pivotal role. The soul, as a direct emanation of the divine, is part of this larger movement of descent and return. While the soul experiences the fragmentation of the material world, it also carries within it the potential for reunion with the divine. The spiritual journey of the soul is, in many ways, a microcosm of the larger cosmic process. As the soul works to purify itself and reconnect with its higher aspects, it participates in the cycle of return, contributing to the restoration of the divine light to the Source.

The process of return, however, is not automatic. It requires conscious effort and spiritual practice. In Neoplatonism, the ascent of the soul is achieved through contemplation and philosophical understanding. The soul must learn to detach itself from the material world and the distractions of the senses, turning inward toward the contemplation of the eternal truths. This process of introspection and purification allows the soul to ascend through the levels of reality, gradually reawakening its connection to the Nous (Divine Intellect) and, ultimately, to the One.

In Gnosticism, the return is facilitated by gnosis—the direct, experiential knowledge of the divine. This knowledge is not intellectual but spiritual, a deep inner awakening that allows the soul to recognize its true nature and its connection to the divine light. Through gnosis, the soul is able to transcend the limitations of the material world and ascend back to the Pleroma. The return to the Pleroma is not just a personal achievement but part of the larger restoration of the divine order, where the scattered light is gathered and returned to its original fullness.

Kabbalistic teachings emphasize the role of human actions in the cycle of return. The soul's journey is intertwined with the practice of tikkun, where each individual has the potential to contribute to the repair of the cosmos. Every positive action, every moment of spiritual awareness, helps to elevate the divine sparks trapped within the material world, bringing them closer to their return to the Source. This process is both individual and collective, where the actions of each soul contribute to the larger cosmic cycle. The ultimate goal is the reunification of all creation with the Infinite Light, where the distinction between the material and spiritual worlds dissolves, and all things return to their original state of unity.

This cyclical process of emanation and return is not confined to the human soul alone. It is a universal principle that governs all of existence. Just as the divine light flows outward from the Source, creating the multiplicity of forms that make up the universe, so too does that light seek to return to its origin. This return is a process of reabsorption, where the multiplicity of creation is gathered back into the unity of the divine. The material world, with all its complexity and diversity, is not separate from the divine but is part of the ongoing cycle of emanation and return.

The return of emanations also speaks to the idea of cosmic balance. The descent of light into the material world, while necessary for the existence of the cosmos, introduces fragmentation and separation. The cycle of return is the process by which this fragmentation is healed, and the original unity of the divine is restored. In this way, the return is not simply a reversal of the process of emanation but a completion of it. The universe is not static but dynamic, constantly moving between the poles of emanation and return, creation and dissolution.

This cyclical movement can also be understood as a reflection of the deeper rhythms of existence. Just as the physical world is governed by cycles—day and night, the changing of seasons, birth and death—so too is the spiritual world governed by cycles of emanation and return. The soul's journey, like the journey of the universe itself, is one of constant movement and transformation. Through this process, the soul not only seeks its own return to the divine but also participates in the larger cosmic dance, where all things flow outward from the Source and eventually return to it.

Ultimately, the cycle of return represents the eternal movement of existence, where the divine light, having descended into the multiplicity of creation, seeks to reunite with its origin. This return is not a negation of creation but a fulfillment of it, where the scattered sparks of divine light are gathered and restored to their original state of unity. The process of emanation and return is the heartbeat of the cosmos, a rhythm that governs all levels of reality, from the highest spiritual realms to the material world. Through this cycle, the universe remains connected to the Source, constantly renewing itself as it moves between the poles of creation and return.

The cycle of return is not merely a metaphysical concept; it is the heartbeat of existence, the pulse that moves through all layers of reality, guiding emanations back to the Source. While the initial phase of the cycle involves the soul's descent into multiplicity and fragmentation, the return is a process of reintegration—a journey through which both the individual soul and the greater cosmos seek to reestablish unity with the Absolute.

The return to the Source begins with an inner awakening. As emanations descend into denser realms of matter, their connection to the divine light becomes obscured. This is particularly true for the human soul, which, clothed in the limitations of the material body, often forgets its divine origin. However, the impulse to return is inherent within all emanations, and this manifests as a deep inner longing—a yearning for something beyond the confines of ordinary existence. This longing is the first step in the cycle of return, a subtle call from the soul's higher aspects that begins to stir the consciousness toward remembrance.

In many spiritual traditions, the journey of return is described as an awakening from illusion. In Gnosticism, this awakening is known as gnosis, a profound and direct knowledge of the divine that allows the soul to see beyond the material world's limitations. Gnosis breaks the spell of ignorance and reveals the true nature of the soul as a spark of the divine light, trapped in a world of shadows. The Gnostic path emphasizes spiritual insight and inner revelation as the keys to unlocking this awareness, enabling the soul to ascend through the layers of reality and return to the Pleroma, the fullness of divine light.

Similarly, in Neoplatonism, the path of return involves the soul's reorientation toward the Nous, or Divine Intellect. The soul, having descended into the material world, becomes distracted by the sensory experiences and desires of earthly life. To return, the soul must turn inward, detaching itself from the distractions of the physical world and focusing on contemplation of the eternal truths. This process of recollection, or anamnesis, is central to the soul's ascent, as it allows the individual to reconnect with the higher aspects of their being and ultimately reunite with the One, the source of all existence.

The role of spiritual practices in facilitating the return is emphasized in many mystical traditions. Practices such as meditation, prayer, contemplation, and ritual are not merely tools for personal growth; they are vehicles for aligning the soul with the divine flow of return. These practices help to purify the soul, stripping away the attachments and illusions that bind it to the material world, and allowing the inner light to shine through. In the Kabbalistic tradition, the act of returning is closely tied to the concept of tikkun, or cosmic repair. Through acts of kindness, spiritual discipline, and the pursuit of wisdom, the soul participates in the process of gathering the scattered sparks of divine light and restoring them to the Source.

Meditation, particularly, plays a crucial role in the return journey. Through meditation, the soul silences the noise of the outer world and turns its attention inward, toward the deeper layers of its own being. In this stillness, the soul begins to perceive the faint echoes of its divine origin, and with sustained practice, these echoes become clearer and more resonant. Mystical traditions across the world have emphasized that through deep meditation, the practitioner can experience a direct encounter with the divine, dissolving the boundaries between the individual self and the Absolute. This state of union, known in many traditions as samadhi, nirvana, or mystical union, is the culmination of the soul's return—a moment when the dualities of the material world fall away, and only the oneness of the divine remains.

Prayer, too, serves as a powerful method of return, particularly when understood as a form of communion rather than petition. In contemplative traditions, prayer is not about asking for divine intervention but about aligning the will of the individual with the will of the divine. Through this form of prayer, the soul begins to harmonize with the higher emanations, allowing the divine light to flow more freely through it. This alignment is a key aspect of the return, as it restores the natural balance between the soul and the Source, facilitating the soul's upward movement through the layers of reality.

Ritual practices, especially those found in mystical and esoteric traditions, also play an essential role in the return. In many cultures, rituals are designed to mimic the cosmic process of emanation and return, using symbolic actions, invocations, and sacred objects to represent the flow of divine energy. These rituals serve as a bridge between the material and spiritual realms, helping participants to realign their consciousness with the higher realities. In this way, ritual becomes a form of spiritual technology, a means of accelerating the soul's return by tapping into the deeper rhythms of the cosmos.

However, the return is not solely dependent on conscious spiritual effort. The cosmic process itself, the very structure of reality, is designed to facilitate the return of all things to the Source. Just as the force of gravity pulls objects toward the center of the Earth, the divine energy that flows through all levels of reality pulls all emanations back toward unity. This natural return process occurs independently of individual effort, guiding all aspects of creation back to their origin. In this sense, even those who are not consciously engaged in spiritual practice are still participating in the cycle of return, as their souls, like all emanations, are naturally drawn back to the Source over time.

In Kabbalistic cosmology, this natural return is expressed through the flow of the sefirot, where divine energy moves in both directions—downward into creation and upward toward the Infinite (Ein Sof). While human beings can accelerate this return through tikkun and spiritual practices, the flow of return is always present, subtly drawing all things back toward unity. The universe, in this sense, is always moving toward greater harmony and reintegration, even when it appears fragmented and chaotic from the perspective of the material world.

Similarly, in Neoplatonism, Plotinus describes the natural inclination of all emanations to seek reunion with the One. The soul, even when immersed in the physical world, retains an inherent pull toward the divine. This inner pull is the source of all spiritual longing and is ultimately what drives the soul to seek out philosophical contemplation and spiritual practice. The return to the One is not simply a personal journey but the fulfillment of the soul's true nature, the completion of a cycle that began with the soul's emanation from the divine.

The return of emanations is, therefore, both a personal and cosmic process. On an individual level, each soul embarks on its own journey of return, working to purify itself, awaken its higher aspects, and reunite with the divine light. On a cosmic level, all of creation participates in the return, as the fragmented aspects of the divine light are gathered and restored to the Source. This dual process ensures that the cycle of emanation and return remains balanced, with the outward flow of creation always accompanied by the inward flow of return.

The cycle of return is also closely tied to the concept of spiritual ascension. In many traditions, the return is described as an ascent through the layers of reality, where the soul moves upward from the densest realms of matter toward the highest realms of spirit. Each level of ascent brings the soul closer to the Source, as it sheds the illusions and limitations of the lower worlds and embraces the light of the higher emanations. This ascent is not only a vertical movement but a process of deepening awareness, where the soul becomes more and more attuned to the divine presence that pervades all things.

Ultimately, the cycle of return is a process of healing and reintegration. The descent of emanations into the material world introduces fragmentation, separation, and illusion, but the return brings all things back into alignment with the original unity. The journey of return is not about escaping the material world but about transforming it, allowing the divine light to shine through all aspects of existence. As each soul returns to the Source, it contributes to the larger process of cosmic repair, helping to restore the harmony of the universe and bring all emanations back into the fullness of the divine light.

Chapter 10
Emanations and Time

Time, as it unfolds in the material universe, is one of the most elusive yet fundamental aspects of existence. When viewed through the lens of the Theory of Emanations, time is not simply a linear sequence of moments. It is an emanation itself—a flow that arises from the deeper layers of spiritual reality, gradually descending through the levels of existence until it crystallizes in the familiar chronological patterns we experience in the material world. Time, in this sense, is a bridge between the eternal and the temporal, and its movement reflects the dynamic process of emanation.

In the highest spiritual realms, time does not exist in the way we understand it. The Source, or Absolute, is beyond time, dwelling in a state of pure eternity where all moments are simultaneously present, unchanging, and unified. In this realm, there is no past or future—only the eternal now. This timelessness is the natural state of the divine, where all emanations exist in their pure, undifferentiated form. It is only when these emanations begin to descend, taking on form and structure, that the concept of time begins to emerge.

In the higher spiritual realms, where emanations are still close to the Source, time is experienced in a non-linear way. Rather than unfolding as a sequence of events, time in these realms is more akin to a vast, interconnected web of possibilities. It is fluid and flexible, allowing for the simultaneous existence of multiple timelines, realities, and experiences. Here, time is not a limitation but a dimension through which spiritual beings can navigate and explore the manifold expressions of divine will. This understanding of time is present in many mystical traditions, which describe higher planes of existence as realms where past, present, and future are not separate but interwoven into a single, unified experience.

As emanations descend further into the lower spiritual realms, time begins to take on a more structured form. The flow of time becomes sequential, though still not as rigid as it appears in the material world. In these realms, time is cyclical rather than strictly linear. The cycles of time—birth, death, renewal—reflect the patterns of emanation and return, where all things move through phases of creation, transformation, and dissolution. These cycles are mirrored in the natural world, in the rhythms of nature, the changing of the seasons, and the cosmic movements of stars and planets.

In Kabbalah, time is understood as an emanation that flows through the sefirot. At its highest levels, time is eternal, connected to the infinite light of Ein Sof. As it descends through the sefirot, time becomes increasingly fragmented, eventually manifesting in the linear, chronological time that governs the material world. Each sefirah influences time in different ways, creating various qualities of time that correspond to different levels of existence. In this sense, time is not a singular, uniform experience but a multi-layered emanation that changes as it moves through the spiritual and material worlds.

One of the key aspects of time in the lower realms is its role in the soul's journey. Time, as it is experienced by human beings, is both a gift and a challenge. It provides the framework within which spiritual growth and transformation can occur, offering the soul opportunities to engage with the material world, learn from its experiences, and work toward its return to the Source. However, the linear nature of time in the material world also imposes limits. It creates a sense of separation from the eternal, making it easy for souls to become trapped in the illusion of time as a fixed, unchanging reality.

In Gnostic thought, time is often viewed as part of the illusion that binds souls to the material world. The Demiurge, the flawed creator of the material universe, is seen as the architect of time, trapping souls in a cycle of birth and death, ignorance and suffering. For the Gnostics, breaking free from the constraints of time is a crucial part of the soul's liberation. Through gnosis, the soul transcends the limitations of temporal existence, awakening to its eternal nature and escaping the cycle of reincarnation and material entrapment.

Neoplatonism, on the other hand, views time as an essential aspect of the soul's journey. Plotinus describes time as the "moving image of eternity"—a reflection of the eternal truths of the Nous, or Divine Intellect, as they unfold in the temporal world. Time, in this view, is not inherently negative but a necessary part of the process of emanation and return. It allows the soul to experience growth, transformation, and purification as it moves through the cycles of life, death, and rebirth. The soul's ultimate goal, however, is to transcend time and return to the timeless state of unity with the One.

The cyclical nature of time is also central to many esoteric traditions. Time is not seen as a straight line leading from creation to dissolution but as a spiral, where the same themes and patterns repeat on increasingly refined levels. These cycles are evident in the movement of the planets, the changing of the seasons, and the stages of human life. They are also reflected in the larger cosmic cycles, where entire worlds and universes move through phases of creation, destruction, and renewal. The idea of cosmic cycles is present in Hinduism's concept of yugas, vast ages of time that mark the rise and fall of civilizations and spiritual awareness. In this sense, time is both a vehicle for spiritual evolution and a reminder of the impermanence of all things.

In the material world, time becomes more rigid and linear. It is experienced as a constant flow from past to future, with the present moment acting as a fleeting point of transition. This linearity is a product of the densification of the divine light as it descends into the material world. The soul, inhabiting a physical body, experiences time as a series of moments, bound by the cycles of birth, growth, decay, and death. This experience of time is often described as a veil that hides the true, eternal nature of reality. Yet, even in the material world, time retains its connection to the higher realms. It is through the experience of time that the soul is able to engage with the process of emanation, work through its karma, and move toward spiritual awakening.

One of the most profound insights offered by the Theory of Emanations is the idea that time itself is a tool for spiritual evolution. In the material world, time provides the structure within which the soul can engage with the challenges and opportunities of earthly existence. It is through the passage of time that the soul learns, grows, and moves closer to the Source. Each moment, each cycle of time, offers the soul a new opportunity to align itself with the divine will and work toward its ultimate return to the eternal.

Time, in this sense, is not something to be escaped or transcended, but embraced as part of the divine plan. The experience of time, with all its limitations and constraints, is a vital aspect of the soul's journey. Through time, the soul comes to understand the nature of impermanence, the ebb and flow of creation and dissolution, and the ultimate unity that lies beyond the dualities of past and future.

Thus, time, as an emanation, is not a barrier but a pathway. It reflects the divine process of creation, where the eternal flows into the temporal, and the temporal, in turn, seeks to return to the eternal. In understanding time through the lens of emanation, we come to see that every moment is infused with the divine, every cycle is part of the greater cosmic rhythm, and every experience of time is an opportunity for the soul to move closer to the Source.

As we deepen our exploration of the nature of time within the context of the Theory of Emanations, it becomes clear that time is not merely a vehicle for spiritual growth, but a crucial aspect of the very fabric of creation itself.

From the perspective of emanation, time is the force that allows the flow of divine energy to unfold in a structured manner. Without time, the process of emanation—from the undifferentiated Source through the layers of existence—would be instantaneous and without sequence. However, time introduces a sense of order to this process, allowing the emanations to unfold gradually, giving rise to the complexity and richness of the created worlds. Time is, therefore, intimately linked to the process of creation, as it permits the gradual transition from the eternal, timeless state of the Absolute into the temporal, differentiated realities we experience.

In the higher spiritual realms, where the emanations remain close to the Source, time is experienced not as a linear progression but as a vast, eternal present. This eternal time, or "sacred time," is often described as a continuous unfolding of divine will, where all moments coexist simultaneously, yet each contains its own depth and meaning. The concept of eternal time is present in many mystical traditions, where the higher planes of existence are seen as realms beyond the limitations of past, present, and future. In these realms, time is a dimension of consciousness rather than a fixed structure, and spiritual beings navigate through it as they would through space, experiencing various states of existence in a fluid, non-linear fashion.

As time descends through the layers of emanation, it begins to take on more defined and structured qualities. In the intermediate spiritual realms, time is experienced as cyclical, reflecting the repetitive patterns of creation and dissolution, birth and death, expansion and contraction. These cycles are present in the rhythms of nature—the changing of the seasons, the phases of the moon, the rise and fall of civilizations—and they echo the larger cosmic cycles that govern the unfolding of the universe. The cyclical nature of time in these realms is often seen as a reflection of the divine process of emanation and return, where the outward flow of energy is always balanced by an inward movement toward reintegration with the Source.

In Kabbalistic thought, time is viewed as a manifestation of the flow of divine energy through the sefirot. Each sefirah represents a different quality of time, from the timelessness of Keter, which is closest to the Infinite (Ein Sof), to the structured, sequential time of Malkhut, which governs the material world. This hierarchy of time reflects the descent of divine light through the layers of reality, with time becoming more fragmented and linear as it moves further from the Source. Yet, even in the material world, time retains its connection to the higher realms, acting as a bridge between the temporal and the eternal. Every moment of time is infused with the divine presence, offering the possibility for spiritual awakening and return.

One of the key insights of the Theory of Emanations is that time, like all other emanations, seeks to return to its origin. The flow of time in the material world is not a one-way journey from creation to dissolution but part of a larger cycle of emanation and return. In this sense, time is not simply a passive backdrop to the events of the universe but an active force that participates in the cosmic process of integration and reintegration. The passage of time, with its cycles of growth, decay, and renewal, mirrors the spiritual journey of the soul as it moves through the stages of existence, working toward its ultimate return to the Source.

This view of time as a tool for spiritual evolution is present in many esoteric traditions. In Hinduism, the concept of yugas—vast cosmic ages that govern the rise and fall of spiritual awareness—reflects the idea that time itself is an evolving force. Each yuga represents a different phase in the cosmic cycle, from the golden age of Satya Yuga, where spiritual truth and harmony prevail, to the darker age of Kali Yuga, where ignorance and materialism dominate. Yet, even in the darkest ages, time continues to move toward renewal, with each cycle offering the opportunity for spiritual evolution and the eventual return to the golden age.

In Neoplatonism, Plotinus describes time as the "moving image of eternity," a reflection of the eternal truths of the Nous (Divine Intellect) as they unfold in the temporal world. Time, in this view, is a necessary condition for the soul's journey through the various levels of existence. It allows the soul to experience growth, learning, and purification as it moves through the cycles of life and death, gradually ascending back toward the eternal state of unity with the One. The cyclical nature of time is not a hindrance to the soul's evolution but a framework within which the soul can work through its attachments and desires, ultimately transcending time and returning to the timeless realm of the divine.

In Gnosticism, time is often portrayed as a trap—a construct of the Demiurge that binds souls to the material world. The linear progression of time, with its cycles of birth, death, and reincarnation, is seen as part of the illusion that keeps souls from realizing their true nature. However, through gnosis—direct knowledge of the divine—the soul can break free from the constraints of time and awaken to its eternal nature. In this sense, time is both a challenge and a pathway: while it creates the illusion of separation, it also offers the possibility for transcendence. Each moment of time holds the potential for spiritual awakening, as the soul moves closer to gnosis and liberation from the material world.

In the material world, time is experienced as a linear sequence, with the past giving rise to the present and the present leading into the future. This linearity is a reflection of the densification of divine light as it descends into the lowest realms of emanation. Time becomes fixed and rigid, creating the illusion of separation between moments and events. However, even in this linear experience of time, the deeper rhythms of the cosmos are still present. The cycles of nature, the movement of the planets, and the stages of human life all reflect the underlying patterns of emanation and return, reminding us that time is not as rigid as it appears.

The soul, as it journeys through time, is given the opportunity to align itself with the higher emanations and work toward its return to the Source. Each moment of time offers a choice: to become more deeply entangled in the material world or to awaken to the divine presence that permeates all of existence. Spiritual practices such as meditation, contemplation, and prayer help to transcend the illusion of linear time, allowing the soul to experience the deeper, cyclical nature of reality and to connect with the eternal. In these practices, the soul steps outside the flow of ordinary time and enters into a state of timelessness, where it can commune directly with the divine.

In Kabbalah, the concept of sacred time is reflected in the practice of observing the Sabbath, which is seen as a moment when the ordinary flow of time is suspended, and the soul is invited to enter into a state of rest and communion with the divine. The Sabbath is a reflection of the eternal now, a glimpse into the timeless realm of the divine, where the soul can experience a moment of union with the Source. This practice reminds us that time is not a linear prison but a tool for spiritual growth, offering moments of transcendence and connection with the higher realms.

Ultimately, the experience of time in the material world is a reflection of the soul's journey through the process of emanation and return. Time is both a guide and a challenge, providing the structure within which the soul can work through its karma, learn from its experiences, and move closer to the Source. The passage of time, with its cycles of growth, decay, and renewal, mirrors the soul's own process of purification and transformation, as it gradually sheds the layers of illusion and awakens to its true nature.

Time, in this sense, is not an obstacle to spiritual enlightenment but a pathway. It is through the experience of time that the soul comes to understand the nature of impermanence, the cycles of creation and dissolution, and the ultimate unity that lies beyond the dualities of past and future. Each moment of time is an opportunity for the soul to awaken to the divine presence within and to move closer to the eternal state of unity with the Source.

As we consider the role of time in the Theory of Emanations, we come to see that time is not a static or linear force but a dynamic, evolving emanation that flows through all levels of reality. It is both a reflection of the divine will and a vehicle for spiritual evolution, offering the soul countless opportunities to realign itself with the higher emanations and to work toward its return to the Source. In this way, time becomes not just a measure of the material world but a sacred rhythm, a cosmic pulse that guides all of creation toward its ultimate reunion with the divine.

Chapter 11
The Role of Evil and Imperfection in Emanations

As we move further into the complexities of the Theory of Emanations, we encounter one of the most challenging concepts: the existence of evil and imperfection within a system that originates from a pure, divine Source. If the universe and all its layers of reality stem from a perfect, infinite Absolute, how can evil, suffering, and imperfection arise within the flow of emanations? The answer to this question is not a simple one, as it touches on the very nature of creation, the process of differentiation from unity, and the inherent limitations of manifestation as emanations move further from their Source.

In many esoteric traditions, evil and imperfection are not seen as independent forces but rather as distortions or diminutions of the divine light as it descends through the layers of reality. The further an emanation moves from the Absolute, the more fragmented and separated it becomes. This separation leads to a weakening of the original light, creating shadows and obscurities within the lower levels of existence. These shadows are not the opposite of the divine but arise from the attenuation of the divine presence. As light fades, darkness emerges—not as an entity in itself but as the absence or reduction of light.

In Gnosticism, this concept is vividly illustrated through the myth of the Demiurge, the lower creator god who shapes the material world. The Demiurge, in this view, is not inherently evil but is ignorant of the higher realms of the Pleroma, the fullness of divine light. He creates the material world in a state of blindness, unaware of the higher truths and spiritual realities that exist beyond his limited scope. The material world, therefore, is flawed not because it is the product of evil, but because it is the work of a being separated from the true light. In Gnosticism, the imperfections of the world reflect this separation from the divine, where the light is dimmed, and shadows, in the form of suffering, ignorance, and limitation, take hold.

Similarly, in Neoplatonism, evil is understood not as a positive force but as the absence or privation of good. Plotinus describes evil as a byproduct of the soul's descent into the material world. As the soul moves further from the Nous, or Divine Intellect, it loses its connection to the eternal truths and becomes entangled in the illusions of the sensory world. In this state of fragmentation, the soul experiences suffering, ignorance, and desire, which are seen as distortions of its true nature. Evil, in this sense, is not an active principle but the result of the soul's forgetfulness of its origin in the divine light. It is a condition of separation, where the soul, cut off from its Source, falls into error and confusion.

In the Kabbalistic tradition, the concept of evil is deeply connected to the process of creation and the structure of the sefirot. As the divine light flows from the Infinite (Ein Sof) through the sefirot, it becomes increasingly filtered and restricted. This restriction is necessary for the creation of the material world, as the raw power of the divine light would overwhelm and dissolve any form of differentiation. However, in the process of this restriction, or tzimtzum, the light can become too restricted, leading to imbalance and the emergence of what is known as klipot—shells or husks that obscure the divine light. These klipot represent the forces of negativity and distortion, but they are not inherently evil. Instead, they are a byproduct of the imbalance created when the flow of divine energy is blocked or diverted. The task of spiritual practice in Kabbalah is to break through these klipot, revealing the hidden light within and restoring balance to the cosmos.

In these traditions, evil and imperfection are not absolute but are seen as relative conditions that arise from the process of emanation itself. As the divine light moves further from its

Source, it becomes increasingly fragmented and separated, giving rise to the potential for imperfection. This fragmentation is necessary for the creation of the material world, where multiplicity and differentiation are key to the richness and diversity of existence. However, this very separation also creates the conditions for suffering, ignorance, and evil, as the beings and forms that inhabit the lower levels of emanation lose touch with the unity of the divine and fall into states of confusion and conflict.

One of the central themes in the exploration of evil and imperfection within emanation is the concept of free will. In the higher realms, where the emanations remain close to the Source, the will of the divine flows freely and harmoniously through all beings. There is no separation between the will of the individual and the will of the divine, as all actions are aligned with the greater cosmic order. However, as emanations descend into the lower realms, particularly in the human experience, free will emerges as a consequence of separation from the divine will. This freedom allows for creativity, growth, and transformation, but it also introduces the possibility of error, misalignment, and disharmony.

In the human experience, free will is both a gift and a challenge. It allows the soul to make choices, to engage with the material world, and to shape its own path of return to the Source. However, this freedom also opens the door to ignorance and suffering, as the soul can choose to turn away from the light and become entangled in the illusions of the material world. In this sense, evil and imperfection are byproducts of the soul's misuse of free will, where the soul, separated from its divine origin, falls into states of selfishness, greed, and violence. These conditions are not the work of an external evil force but are the natural consequences of the soul's disconnection from the divine.

The existence of evil and imperfection within the framework of emanation also serves a deeper spiritual purpose. While these conditions are painful and often tragic, they are not meaningless. Many traditions, particularly Kabbalah, view the presence of evil as a necessary part of the cosmic process of tikkun, or repair. The imperfections of the world, the suffering and darkness that arise from the separation of light, are opportunities for healing and transformation. Through acts of kindness, compassion, and spiritual awareness, the divine light that is hidden within the darkness can be revealed and restored to its original state. In this sense, evil is not an obstacle to be destroyed but a force to be transformed, integrated into the greater harmony of creation.

This process of transformation is mirrored in the individual soul's journey. The soul, in its descent into the material world, encounters suffering, ignorance, and temptation. These experiences, while painful, are also opportunities for growth and purification. By confronting and overcoming the forces of darkness within and without, the soul strengthens its connection to the divine light and moves closer to its ultimate return to the Source. The journey through imperfection, therefore, is not a deviation from the divine plan but an integral part of it, where the soul learns to transcend its limitations and realign itself with the higher emanations.

In many ways, the presence of evil and imperfection in the process of emanation reflects the complexity of creation itself. The divine light, in its infinite wisdom, allows for the emergence of multiplicity, differentiation, and free will. These conditions create the richness and diversity of existence, but they also introduce the possibility of fragmentation and distortion. Yet, even in the darkest corners of the universe, the light of the Source is never entirely lost. It remains hidden, waiting to be revealed and restored through the efforts of spiritual beings, human souls, and the larger cosmic process of emanation and return.

Thus, evil and imperfection, while difficult to comprehend, are woven into the very fabric of creation. They are the shadows cast by the divine light as it descends into the lower realms, reflections of the separation and fragmentation that are inherent in the process of manifestation. Yet, these shadows also serve as catalysts for spiritual growth and cosmic healing, offering opportunities for transformation, redemption, and the eventual restoration of unity.

As we delve further into the exploration of evil and imperfection within the structure of emanation, we move from understanding their origins and manifestations to considering how they are ultimately part of the cosmic process of redemption and transformation. Just as the descent of divine light into the lower worlds inevitably creates shadows and fragmentation, the return of this light to the Source involves the gradual purification, transmutation, and integration of these shadows.

The idea of redemption in the context of the Theory of Emanations is multifaceted. It is not simply about the vanquishing of evil or the elimination of imperfection, but about the healing and restoration of balance within the cosmic order. The imperfections that arise in the lower realms of existence, particularly within the material world, are opportunities for the divine light to be rediscovered, elevated, and reintegrated into the greater harmony of creation. This process of redemption is not limited to the individual soul but encompasses the entire cosmos, as every aspect of creation participates in the movement toward reintegration with the Source.

In Kabbalistic tradition, this process is encapsulated in the concept of tikkun olam, the "repair of the world." The idea of tikkun suggests that the universe, while seemingly fractured and imperfect, is in a constant state of restoration. The divine light, which becomes hidden or fragmented as it descends through the layers of reality, is scattered throughout the material world, trapped within the klipot—the shells or husks that obscure its brilliance. The task of spiritual beings, and particularly human souls, is to engage in acts of repair, freeing the divine sparks from their concealment and elevating them back toward the Source. Every act of kindness, every moment of spiritual awareness, contributes to this cosmic repair, helping to restore the balance between light and darkness and bringing the universe closer to its original unity.

The redemption of evil and imperfection is not only a cosmic process but also an individual one. Each soul, in its journey through the material world, carries with it the potential for both light and shadow. The challenges and struggles that arise in life—suffering, ignorance, temptation—are not random but are integral to the soul's process of growth and evolution. These challenges, often experienced as manifestations of evil or imperfection, are opportunities for the soul to confront its limitations, transform its attachments, and move closer to its divine nature. In this way, the darkness that the soul encounters is not an obstacle to be feared but a necessary part of the journey toward spiritual awakening.

In Gnostic traditions, the process of redemption is closely tied to the concept of gnosis—direct, experiential knowledge of the divine. The material world, governed by the Demiurge and his Archons, is seen as a realm of ignorance and illusion, where the soul is trapped in a cycle of suffering and forgetfulness. Yet, within this realm of darkness, the soul carries a spark of the divine light, a fragment of the Pleroma. The Gnostic path involves awakening this inner light through gnosis, transcending the illusions of the material world, and returning to the fullness of the divine. This awakening is not merely an escape from the world but a transformation of the self, where the soul overcomes its entanglement with the lower emanations and reclaims its true nature as a being of light.

Similarly, in Neoplatonism, the redemption of the soul involves a process of purification and ascent. Plotinus describes how the soul, in its descent into the material world, becomes mired in the illusions and desires of the sensory realm, losing its connection to the higher realities of the Nous (Divine Intellect). The return of the soul to the One requires a process of inner purification, where the soul sheds the attachments and distractions of the lower world and reorients itself toward the eternal truths of the higher realms. This ascent is a gradual process of transformation, where the soul moves from fragmentation and separation back toward unity and wholeness.

The notion of transmutation plays a significant role in many mystical traditions, particularly in alchemy, which views the transformation of base materials into gold as a metaphor for the spiritual process of redeeming the soul from its fallen state. The alchemist's work is not just about turning lead into gold but about transforming the imperfections of the material world into something pure and divine. In this context, the "base materials" represent the lower aspects of existence—ignorance, desire, and suffering—while the "gold" symbolizes the purified soul, which has been freed from its attachments and reconnected with the divine light. This process of alchemical transformation mirrors the larger cosmic process of redeeming the fallen world, where even the darkest elements of creation are ultimately transmuted into light.

One of the key insights of the Theory of Emanations is that even evil and imperfection have a role to play in the grand scheme of creation. They are not external forces that oppose the divine but are part of the natural consequence of differentiation and separation from the Source. In the lower realms, where light becomes fragmented, these shadows emerge as a byproduct of the emanation process. However, their existence is not without purpose. The presence of evil and imperfection creates the conditions necessary for free will, growth, and spiritual evolution. It is through confronting and transforming these forces that the soul learns to transcend its limitations and move closer to the divine.

In this sense, evil and imperfection are not permanent states but are fluid and mutable, subject to change through the process of spiritual evolution. The forces of darkness that arise in the lower worlds are not fixed but are always in the process of being transformed, either by individual souls working through their own challenges or by the larger cosmic process of redemption. This dynamic nature of evil and imperfection is central to the idea of tikkun, where the broken and fragmented aspects of creation are constantly being repaired and reintegrated into the whole.

The methods for redeeming and transmuting evil are found in the spiritual practices of many traditions. Meditation, prayer, contemplation, and acts of compassion all serve to realign the soul with the higher emanations and bring light into the darkness. These practices are not just about personal enlightenment but are part of the larger cosmic process of healing and restoration. Every act of spiritual awareness, every moment of love and kindness, contributes to the elevation of the divine sparks hidden within the material world, helping to dissolve the klipot and reveal the underlying unity of creation.

Ritual and symbolic actions also play a significant role in the redemption of evil. In Kabbalistic rituals, for example, the act of lighting candles or reciting prayers is seen as a way of drawing down divine light into the material world, illuminating the darkness and helping to restore balance. These rituals are not merely symbolic but are understood as direct interventions in the cosmic process of tikkun, where the practitioner participates in the ongoing work of repairing the world and elevating the divine sparks.

The presence of evil and imperfection in the world is, therefore, not something to be feared or avoided but to be understood as part of the larger process of emanation and return. These forces, while painful and challenging, are opportunities for transformation and healing. They serve as catalysts for spiritual growth, pushing the soul to confront its limitations and to seek out the light within itself. Through the process of redemption, even the darkest aspects of creation can be reintegrated into the divine, bringing the universe one step closer to its ultimate state of unity.

Ultimately, the role of evil and imperfection in the Theory of Emanations reflects the complexity and depth of the creative process. While the descent of the divine light into the material world inevitably creates shadows and fragmentation, these shadows are not separate from the light but are part of the same process of differentiation. The journey of emanation, with all its challenges, is also a journey of return, where the forces of darkness and imperfection are gradually transformed and reintegrated into the greater harmony of creation. In this way, the presence of evil is not a contradiction but a necessary aspect of the universe's evolution toward its final, redeemed state.

Chapter 12
The Process of Returning to the Source

The journey of emanation is, by its very nature, twofold. As emanations flow from the Absolute, descending through the layers of existence, the return process is simultaneously set into motion. This return, known in many traditions as the spiritual path, is the ascent back to the Source—a movement of reintegration and reunification with the divine. The descent into multiplicity and separation is followed by an upward movement, where the soul and all created forms seek to recover their lost unity.

The return to the Source is not simply a reversal of the emanation process but a conscious, intentional journey. In the early stages of emanation, the flow of divine energy moves outward effortlessly, creating layers of existence through a natural, continuous process. However, the return requires an active participation, especially in the case of the human soul, which possesses free will and self-awareness. While all beings are naturally drawn toward reunion with the Source, human beings must consciously engage in this return, navigating the challenges of the material world and transcending the illusions of separation.

The first step in the process of return is the awakening of the soul to its true nature. This awakening is often described as a moment of realization, where the soul becomes aware of its divine origin and the path that lies before it. In many spiritual traditions, this awakening is seen as the soul's recognition of the divine spark within itself—a recognition that initiates the journey of return. The soul, which had become immersed in the material world and distracted by its desires and attachments, begins to remember its higher purpose and the ultimate goal of reintegration with the Absolute.

In Neoplatonism, this awakening is described as the soul's recollection of the eternal truths of the Nous, or Divine Intellect. Plotinus emphasizes that the soul, even in its descent into the material world, never fully loses its connection to the higher realms. The knowledge of the divine is always present within the soul, though it may be obscured by the distractions of the sensory world. The process of return, therefore, begins with an inward turning, where the soul detaches itself from the external world and reconnects with the inner light of the Nous. This inward focus is essential for the soul's ascent, as it allows the individual to perceive the higher realities that transcend the limitations of the material world.

In Gnosticism, the awakening of the soul is a dramatic event, often portrayed as a sudden and profound realization of the soul's imprisonment in the material world. The Gnostic path emphasizes the importance of gnosis—direct, experiential knowledge of the divine—as the key to liberation. Through gnosis, the soul awakens to the truth that the material world, created by the Demiurge, is a realm of illusion and separation. This realization initiates the soul's journey of return, as it seeks to escape the confines of the material world and ascend back to the Pleroma, the fullness of divine light. In this sense, the Gnostic awakening is both an individual and cosmic event, as the liberation of the soul contributes to the larger process of restoring the divine order.

The Kabbalistic tradition also places a strong emphasis on the process of awakening and return. The soul, in Kabbalah, is seen as a direct emanation of the divine, descending through the sefirot and eventually incarnating in the material world. However, this descent is not a fall from grace but a necessary part of the soul's journey toward spiritual maturation. The soul must experience the challenges and limitations of the material world in order to develop the qualities needed for its return. The return is facilitated by the practice of tikkun, where the soul actively participates in the repair and elevation of the divine sparks scattered throughout creation. Through acts of kindness, spiritual practice, and the pursuit of wisdom, the soul contributes to the larger cosmic process of healing and reintegration, gradually ascending back through the sefirot toward its original unity with the Source.

One of the central themes of the return process is the idea of purification. As the soul descends into the material world, it becomes entangled in the illusions of separation, desire, and attachment. These entanglements create layers of forgetfulness, obscuring the soul's connection to the divine. The journey of return, therefore, requires a process of purification, where the soul gradually sheds these layers and reawakens to its true nature. This purification is often described as a process of spiritual alchemy, where the soul undergoes a transformation from its base, material state to its higher, divine form.

In many mystical traditions, this process of purification is facilitated by specific spiritual practices. Meditation, prayer, contemplation, and ritual are all tools that help the soul detach from the distractions of the material world and realign itself with the higher emanations. These practices serve to quiet the mind, open the heart, and reconnect the individual with the deeper layers of reality. In Neoplatonism, for example, the practice of philosophical contemplation is seen as a key method for purifying the soul. By contemplating the eternal truths of the Nous, the soul gradually disentangles itself from the sensory world and ascends toward the One, the ultimate source of all existence.

In Gnosticism, the process of purification involves the rejection of the material world and the pursuit of gnosis. The soul, trapped in the realm of illusion, must turn away from the distractions of the flesh and seek the inner light of the divine. This purification is not simply about moral or ethical behavior but involves a deep, inner transformation, where the soul transcends the dualities of good and evil and awakens to its original state of unity with the divine. In this sense, purification is both a process of spiritual discipline and a moment of radical transformation, where the soul is reborn into the light of the Pleroma.

In Kabbalah, the process of purification is closely tied to the concept of teshuvah, or return. Teshuvah is often translated as repentance, but its deeper meaning is the return of the soul to its original state of purity and unity with the divine. Through the practice of teshuvah, the soul recognizes its mistakes and misalignments and seeks to realign itself with the divine will. This process is not about punishment or judgment but about healing and restoration. The soul, through acts of compassion, forgiveness, and spiritual awareness, participates in the larger cosmic process of tikkun, gradually repairing the fractures in the universe and returning to its original state of harmony with the Source.

The journey of return also involves the concept of ascent through the levels of reality. In the same way that the soul descends through layers of emanation in its journey away from the Source, the return involves an ascent back through these layers. This ascent is not merely a physical movement but a process of expanding consciousness, where the soul awakens to higher and higher levels of reality. Each stage of the ascent brings the soul closer to the divine, as it moves from the dense, fragmented realms of the material world toward the pure, undivided light of the Absolute.

In Neoplatonism, this ascent is described as the soul's movement through the stages of purification, illumination, and union. The soul, having purified itself through contemplation and philosophical inquiry, begins to ascend through the levels of reality, shedding the illusions of the lower world and awakening to the higher truths of the Nous. This ascent culminates in the experience of union with the One, where the soul transcends all duality and experiences the oneness of the divine. This state of union is not a loss of individuality but a realization of the soul's true nature as part of the infinite, eternal Source.

In Gnosticism, the ascent is portrayed as the soul's escape from the material world and its return to the Pleroma. The soul, having awakened through gnosis, begins to ascend through the layers of reality, passing through the realms of the Archons and overcoming the obstacles of the material world. This ascent is a journey of liberation, where the soul sheds the false identities imposed by the Demiurge and reclaims its true nature as a being of light. The ultimate goal of this ascent is the return to the Pleroma, where the soul is reunited with the fullness of the divine light and experiences its original state of unity.

In Kabbalah, the ascent of the soul is described as the movement through the sefirot, where the soul gradually elevates itself from the lower realms of material existence toward the higher realms of divine light. This ascent is facilitated by the practice of tikkun, where the soul actively participates in the repair of the universe. Each act of kindness, each moment of spiritual awareness, contributes to the soul's ascent, as it gathers the scattered sparks of divine light and returns them to their original state of unity. This ascent is not only an individual journey but a cosmic process, where the elevation of the soul contributes to the larger movement of the universe toward reintegration with the Source.

The return to the Source, therefore, is not a singular event but a dynamic, ongoing process. It involves both the individual soul's journey of purification and ascent and the larger cosmic process of healing and reintegration. Each soul, through its own efforts and spiritual practice, contributes to the restoration of the divine order, helping to repair the fractures in the universe and bring all things back into harmony with the Absolute. This journey of return is the fulfillment of the soul's original purpose, where it moves from separation and fragmentation back to its original state of unity with the Source.

The first stage of this return is often described as purification, a necessary preparation for the soul's ascent. As the soul becomes aware of its divine origin and begins the process of returning, it must cleanse itself of the attachments and illusions that it has gathered during its descent into the material world. These attachments—desires, fears, and false identities—create barriers between the soul and the higher realms of existence. In many traditions, this purification is not merely a moral process but a metaphysical one, where the soul realigns itself with the flow of divine energy and sheds the layers of separation that obscure its true nature.

In mystical traditions like Sufism, this purification is described as the tariqa, the spiritual path of cleansing the self (or nafs). The journey begins with the soul recognizing its lower nature and working to transcend it through spiritual practices such as prayer, fasting, and remembrance of God (dhikr). As the soul purifies itself, it moves through different stages of consciousness, gradually shedding the ego and experiencing deeper levels of union with the divine. The final goal is fana, the annihilation of the self in the presence of God, where the soul transcends all sense of separation and becomes fully united with the divine will.

In Kabbalah, the process of purification is intimately connected to the practice of tikkun, the repair of the world. As the soul engages in acts of spiritual elevation—through meditation, study of sacred texts, and ethical behavior—it not only purifies itself but also contributes to the healing of the cosmos. Each act of repair helps to realign the fragmented aspects of creation with the divine light, allowing the soul to ascend through the sefirot and reconnect with its higher aspects. This process is seen as both individual and collective, where the elevation of the soul contributes to the elevation of the universe itself.

Once the soul has undergone purification, the next stage of the return is illumination. This is the phase in which the soul begins to perceive the higher realities of existence and experiences direct communion with the divine. In Neoplatonism, Plotinus describes this stage as the awakening of the soul's intellect, where it begins to perceive the eternal truths of the Nous (Divine Intellect). The soul, having turned away from the distractions of the material world, now turns inward, contemplating the forms and ideas that reflect the divine order. This illumination is not merely intellectual but experiential—an inner vision of the divine reality that transcends ordinary perception.

In Gnosticism, this stage of illumination is achieved through gnosis, the direct and experiential knowledge of the divine. The soul, having awakened from the illusions of the material world, now begins to perceive the hidden truths of the spiritual realms. This gnosis allows the soul to see beyond the false constructs of the Demiurge and his Archons, recognizing its true nature as a being of light. The experience of gnosis is transformative, as it not only reveals the higher realities of existence but also elevates the soul, bringing it closer to the fullness of the Pleroma, the realm of divine light.

In Sufism, the stage of illumination is known as ma'rifa—the deep, inner knowledge of God. This knowledge is not intellectual but mystical, a direct encounter with the divine presence that transcends all concepts and dualities. The soul, having purified itself, now experiences moments of divine revelation, where the barriers between the self and God dissolve, and the soul is flooded with the light of divine truth. These moments of illumination are often described as encounters with the "Beloved," where the soul, overwhelmed by love, is drawn deeper into the mystery of divine union.

As the soul continues its ascent, it enters the final stage of return: union with the Source. This union is the culmination of the soul's journey, where it transcends all duality and merges fully with the divine. In Neoplatonism, this union is described as the soul's return to the One, the ultimate source of all existence. At this stage, the soul no longer perceives itself as separate from the divine but recognizes its oneness with the Absolute. This experience of union is beyond words or concepts—it is a state of pure being, where the soul dissolves into the infinite, eternal presence of the One.

In Sufism, this state of union is known as baqa—the state of subsistence in God after the experience of fana (the annihilation of the self). In baqa, the soul, having been annihilated in the presence of the divine, now subsists in a state of perpetual union with God. This is not a static state but a dynamic one, where the soul continually experiences the flow of divine love and presence. The Sufi mystic, having reached this stage, becomes a channel for divine love and wisdom, living in the world but no longer of it, fully aligned with the will of the divine.

In Kabbalah, the union with the Source is symbolized by the reunion of the sefirot with the Infinite (Ein Sof). The soul, having ascended through the levels of the sefirot, now merges with the divine light, experiencing a state of oneness with the Infinite. This union is not a loss of individuality but a realization of the soul's true nature as an emanation of the divine. In this state, the soul participates in the continuous flow of creation and return, fully integrated into the cosmic process of emanation and reintegration.

The return to the Source is, therefore, not the end of the soul's journey but a new beginning. Having reached union with the divine, the soul now participates in the ongoing cycle of creation, acting as a vessel for the divine will. In many mystical traditions, those who achieve union with the Source are seen as enlightened beings or saints, who continue to serve the world, helping others on their own journey of return. These enlightened beings, having transcended the limitations of the material world, now act as bridges between the divine and the human, guiding others toward the light.

The process of return is also deeply connected to the concept of cosmic time. In many traditions, time is seen as a spiral or cycle, where the movement of creation and return is not linear but cyclical. The soul, in its journey of return, moves through these cycles, experiencing the flow of time in increasingly subtle and refined ways. As the soul ascends, it transcends the linear experience of time, entering into the eternal now, where all moments are present in the divine. This experience of timelessness is a key aspect of the soul's union with the Source, where it participates in the eternal flow of divine creation and return.

In some esoteric traditions, the return to the Source is also seen as a return to the "primordial state," the original condition of unity before the fragmentation of emanation. This primordial state is not simply a return to the past but a return to the original potential of the soul, where it fully realizes its divine nature and purpose. The soul, having journeyed through the cycles of creation and dissolution, now returns to its original state of purity and unity, fully integrated into the divine order.

The final union with the Source is, therefore, both a personal and cosmic event. On a personal level, it represents the fulfillment of the soul's deepest longing for reunion with the divine. On a cosmic level, it reflects the larger process of emanation and return, where all of creation is gradually drawn back into the unity of the divine. This union is not a final, static state but an ongoing process, where the soul continues to participate in the flow of divine energy, acting as a conduit for the light of the Source.

In this way, the process of returning to the Source is both a journey and a homecoming. The soul, having traveled through the cycles of existence, returns to its original state of unity, fully realizing its divine nature and purpose. This return is the culmination of the soul's journey, where it merges with the infinite, eternal presence of the Source, experiencing the ultimate fulfillment of its existence. Yet, even in this state of union, the journey continues, as the soul participates in the ongoing flow of creation and return, forever aligned with the divine will.

Chapter 13
The Role of Consciousness in Emanations

The process of emanation, by which all things flow outward from the Source and eventually return to it, is deeply intertwined with the nature of consciousness. Consciousness acts as both the vehicle and the witness of this cosmic journey. In many traditions, consciousness itself is understood as an emanation, originating from the infinite, undivided awareness of the Absolute and progressively descending through various levels of existence. As consciousness moves further from its Source, it becomes increasingly fragmented, but its core remains a reflection of the divine.

At its highest level, consciousness is pure, undifferentiated awareness. This is the state of the Absolute, where there is no distinction between subject and object, knower and known. In this primordial state, consciousness is not something that is possessed or experienced; it is the very essence of existence itself. It is infinite and all-encompassing, beyond time, space, and form. This state is often described as the divine mind, or Nous in Neoplatonism, the source of all intellectual and spiritual realities. At this level, consciousness is inseparable from the One, and all knowledge is immediate, direct, and complete.

As consciousness descends through the layers of emanation, it begins to take on form and structure. In the higher spiritual realms, consciousness remains expansive and unified, but it starts to differentiate itself into various modes of perception and awareness. These higher levels of consciousness are often described as the realms of angels, archangels, or divine beings—entities that exist in a state of constant communion with the Source, but who also possess distinct identities and roles within the cosmic order. These beings reflect different aspects of the divine mind, and their consciousness is still deeply connected to the flow of divine energy.

In Kabbalah, this higher consciousness is represented by the upper sefirot—Keter (Crown), Chokhmah (Wisdom), and Binah (Understanding). These sefirot represent the highest levels of consciousness, where the divine will and intellect are still closely aligned with the Infinite (Ein Sof). In these realms, consciousness is vast and all-encompassing, but it has begun to differentiate into the distinct aspects of divine wisdom and understanding. Here, consciousness perceives the universe as an integrated whole, where all things are interconnected and part of the divine plan. This is the realm of archetypes, where the fundamental patterns of existence are perceived in their purest form.

As consciousness continues its descent, it becomes increasingly fragmented. By the time it reaches the lower spiritual realms and the material world, consciousness has become divided into individual selves—each experiencing a limited and subjective reality. In the material world, consciousness is bound by the limitations of the body, the mind, and the sensory world. It becomes identified with the ego, the personal self, which perceives itself as separate from others and from the divine. This state of fragmentation is the source of much of the suffering and ignorance that characterizes the human experience.

In Gnosticism, this fragmentation of consciousness is viewed as the soul's fall into the material world. The soul, which originally existed in a state of pure, undivided consciousness, becomes trapped in the realm of the Demiurge, where it experiences separation, limitation, and ignorance. The soul forgets its divine origin and becomes entangled in the illusions of the material world. This fragmented consciousness is described as the "sleep" of the soul, where it is unaware of its true nature and its connection to the divine light. The process of awakening, or gnosis, is the reawakening of consciousness to its original state of unity and knowledge.

The role of consciousness in the human experience is complex, as it reflects both the soul's connection to the divine and its entanglement in the material world. On one hand, consciousness is the soul's link to the higher realms, the spark of divine light that guides it on its journey of return. On the other hand, consciousness is also the site of the soul's greatest challenges, as it struggles with the illusions and distractions of the ego and the sensory world. The dual nature of consciousness—its potential for both awakening and entrapment—makes it the key battleground in the soul's journey of return.

In Neoplatonism, the ascent of the soul is closely tied to the expansion of consciousness. Plotinus describes how the soul, in its descent into the material world, becomes enmeshed in the lower faculties of sense perception and desire, losing its connection to the higher realms of the Nous. The return to the One requires the soul to purify its consciousness, detaching itself from the distractions of the material world and turning inward toward the contemplation of eternal truths. This purification of consciousness allows the soul to ascend through the levels of existence, reconnecting with the divine intellect and ultimately merging with the One.

In the Kabbalistic framework, consciousness is similarly viewed as the key to spiritual ascent. The soul, as it descends through the sefirot, experiences increasing fragmentation and limitation. However, through spiritual practice, the soul can expand its consciousness, reconnecting with the higher sefirot and the divine light. This expansion of consciousness is often described as a return to the "higher self," where the individual soul becomes aware of its true nature as an emanation of the divine. Through meditation, prayer, and the study of sacred texts, the soul gradually purifies its consciousness and ascends through the sefirot, reconnecting with the wisdom and understanding of the higher realms.

The expansion of consciousness is not only a personal journey but also a cosmic one. In many traditions, the evolution of consciousness is seen as part of the larger process of cosmic evolution, where the entire universe is gradually awakening to its divine origin. This idea is present in the concept of the Great Chain of Being, where all levels of existence, from the lowest material forms to the highest spiritual realities, are connected in a continuous hierarchy of consciousness. Each level reflects a different aspect of the divine, and as consciousness evolves, it moves closer to the Source.

In esoteric traditions, consciousness is often described as a bridge between the material and spiritual worlds. It is the medium through which the soul experiences both the limitations of the lower realms and the freedom of the higher realms. Through the expansion of consciousness, the soul is able to transcend the illusions of the material world and perceive the deeper truths of existence. This is not a purely intellectual process but a profound transformation of being, where the soul becomes aligned with the divine will and participates in the flow of emanation and return.

One of the most important aspects of consciousness in the process of emanation is its ability to reflect the divine. In its purest form, consciousness is a mirror of the divine mind, reflecting the infinite wisdom and love of the Source. However, as consciousness descends through the layers of existence, this reflection becomes increasingly distorted. In the material world, consciousness is often clouded by ignorance, fear, and desire, obscuring its ability to perceive the divine light. The process of spiritual awakening, therefore, involves the clearing of these distortions, allowing consciousness to once again reflect the divine in its fullness.

In this way, consciousness plays a dual role in the process of emanation. On one hand, it is the means by which the soul experiences separation and limitation. On the other hand, it is the key to the soul's return, as it provides the awareness and insight necessary for the soul to navigate the challenges of the material world and ascend back to the Source. The expansion of consciousness is both the goal and the path of the soul's journey, as it moves from the darkness of ignorance to the light of divine knowledge.

In the human experience, consciousness is a paradoxical force. It is both the source of the soul's suffering and the means of its liberation. As the soul becomes more aware of its divine nature, its consciousness expands, allowing it to perceive the deeper realities of existence. This expansion is not a rejection of the material world but a transformation of perception, where the soul sees through the illusions of separation and experiences the unity of all things. Through this expanded consciousness, the soul begins to participate in the larger process of emanation and return, acting as a channel for the divine light and contributing to the healing and restoration of the universe.

Ultimately, the role of consciousness in the process of emanation is one of alignment. Consciousness is the medium through which the divine will is expressed in the world, and the soul's task is to align its consciousness with the higher emanations. This alignment is not a static state but an ongoing process of purification, expansion, and transformation. As the soul aligns its consciousness with the divine, it participates in the cosmic process of return, gradually moving closer to the Source and contributing to the reintegration of all things into the unity of the divine.

The awakening of consciousness begins with a shift in perception—a recognition that the reality perceived by the senses is not the full extent of existence. In many mystical traditions, this awakening is described as the "opening of the inner eye" or the realization that there are higher planes of reality beyond the material world. This moment of awakening is often accompanied by a profound sense of clarity, as the individual becomes aware of the interconnectedness of all things and the presence of the divine within and around them. This is the first step on the path to enlightenment, where the soul begins to detach from the illusions of separation and moves toward unity with the Source.

In Gnosticism, the awakening of consciousness is understood as gnosis—direct, experiential knowledge of the divine. The soul, trapped in the material world and unaware of its true nature, experiences a "fall" into ignorance. However, through gnosis, the soul can pierce the veil of illusion and recognize its origin in the divine light. This knowledge is not intellectual but is felt as an inner transformation, where the soul awakens to its higher purpose and begins the process of returning to the Pleroma, the realm of fullness and light. Gnosis acts as a catalyst for spiritual evolution, providing the soul with the insight needed to navigate the challenges of the lower realms and ascend back to the Source.

In the Kabbalistic tradition, the expansion of consciousness is tied to the soul's journey through the sefirot. As the soul becomes aware of the divine light within itself, it begins to ascend through the levels of the sefirot, moving from the material realm of Malkhut (Kingdom) to the higher realms of wisdom, understanding, and beyond. This ascent is facilitated by practices such as meditation, prayer, and study of sacred texts, which help the soul align its consciousness with the higher emanations. Each level of the sefirot represents a different aspect of divine consciousness, and as the soul ascends, it gradually reunites with the divine mind, experiencing greater unity and clarity with each step.

The expansion of consciousness is not a passive process—it requires active engagement with the spiritual path. In many traditions, this engagement involves practices that help to purify the mind, quiet the distractions of the material world, and open the heart to the presence of the divine. Meditation is one of the most common methods for expanding consciousness, as it allows the individual to transcend the sensory world and enter into a state of inner stillness, where they can perceive the deeper truths of existence. In this state, the mind becomes a mirror for the divine, reflecting the light of the higher emanations and facilitating the soul's ascent.

In Sufism, the expansion of consciousness is described as the journey of the heart. The heart is seen as the center of spiritual awareness, and through practices such as dhikr (remembrance of God) and muraqaba (meditation), the Sufi mystic seeks to purify the heart and align it with the divine will. As the heart is purified, the mystic experiences deeper levels of union with the divine, moving through stages of spiritual realization that culminate in fana (the annihilation of the self) and baqa (subsistence in God). The expansion of consciousness in Sufism is not just an intellectual awakening but a profound transformation of the self, where the individual becomes a vessel for the divine presence.

The awakening of consciousness also involves the dissolution of the ego. The ego, or the sense of a separate self, is one of the greatest obstacles to spiritual evolution. It creates the illusion of separation between the individual and the divine, leading to attachments, desires, and fears that keep the soul bound to the material world. In order to expand consciousness, the individual must transcend the ego, recognizing that their true self is not separate from the Source but is an expression of the divine light. This process of ego dissolution is often challenging, as it involves letting go of deeply ingrained patterns of thought and behavior. However, it is through this dissolution that the soul can experience the fullness of divine consciousness.

In Neoplatonism, Plotinus describes the process of expanding consciousness as the soul's ascent through the levels of reality, from the sensory world to the intellectual realm of the Nous (Divine Intellect), and ultimately to the One. This ascent requires the soul to detach from the distractions of the material world and turn inward, contemplating the eternal truths that lie beyond the realm of appearances. As the soul ascends, its consciousness expands, allowing it to perceive the unity and harmony of the divine order. The final stage of this ascent is union with the One, where the soul transcends all duality and experiences the infinite, undivided consciousness of the divine.

The concept of enlightenment is closely tied to the expansion of consciousness. In many traditions, enlightenment is described as the state of full awareness, where the individual realizes their unity with the divine and experiences the totality of existence in a single moment. This state is often referred to as satori in Zen Buddhism, moksha in Hinduism, or nirvana in Buddhism. While the terminology may differ, the underlying experience is the same: the individual transcends the limitations of the ego and the material world, merging with the infinite consciousness of the Source.

In the context of the Theory of Emanations, enlightenment represents the culmination of the soul's journey of return. It is the point at which the soul, having expanded its consciousness and purified itself of all attachments, merges fully with the higher emanations and reunites with the Source. This state of union is not a static condition but a dynamic process, where the individual continues to participate in the flow of divine energy, acting as a conduit for the light of the Source in the world. Enlightened beings are often described as living in a state of perpetual awareness, where they are fully present in each moment, yet simultaneously connected to the infinite and eternal reality of the divine.

One of the key insights of the Theory of Emanations is that the expansion of consciousness is not limited to individuals but is part of a larger cosmic process. As individual souls awaken and expand their consciousness, they contribute to the awakening of the entire cosmos. This idea is reflected in the concept of the Great Chain of Being, where all levels of existence are connected in a continuous hierarchy of consciousness. Each level reflects a different aspect of the divine, and as consciousness evolves, it moves closer to the Source. The expansion of consciousness is, therefore, not just a personal journey but a cosmic one, where all of creation is gradually awakening to its divine origin.

The expansion of consciousness is also linked to the idea of spiritual freedom. As consciousness expands, the soul becomes less bound by the limitations of the material world and the ego, experiencing greater levels of freedom and autonomy. This freedom is not simply the ability to make choices but the ability to align one's will with the divine will. In this state of expanded consciousness, the individual experiences a sense of flow, where their actions are guided by the higher emanations, and they are no longer constrained by the fears and desires of the ego. This is the true meaning of spiritual freedom—freedom from the illusions of separation and the limitations of the material world.

Ultimately, the role of consciousness in emanations is one of awakening and alignment. Consciousness is both the vehicle and the guide for the soul's return to the Source. Through the expansion of consciousness, the individual soul becomes aware of its divine nature, transcends the limitations of the ego, and reunites with the higher emanations. This process is not a linear progression but a dynamic journey, where the soul continuously expands its awareness, participates in the flow of divine energy, and contributes to the awakening of the entire cosmos. The expansion of consciousness is, therefore, both the path and the goal of the spiritual journey, leading the soul ever closer to the infinite and eternal reality of the Source.

Chapter 14
The Transformation of Emanations in the Lower Worlds

As emanations descend from the Source, they pass through various layers of reality, gradually losing the purity and unity of their origin. By the time these emanations reach the lower worlds—such as the physical and material planes—they undergo significant transformations.

At the heart of this transformation lies the principle of densification. As divine light moves further from the Source, it becomes more concentrated and condensed. In the higher spiritual realms, emanations remain relatively close to the divine, still possessing clarity and fluidity. However, as they descend through the layers of existence, they begin to take on form, structure, and eventually, materiality. This process of densification is necessary for the manifestation of the physical world but also introduces fragmentation and limitation. The divine light, once unified and infinite, becomes divided into distinct forms, objects, and experiences.

In Kabbalah, the process of densification is reflected in the concept of tzimtzum, the divine contraction that allows the infinite light of the Source to create space for the material world. Through tzimtzum, the divine light is restricted and diminished, enabling the creation of finite realities. This contraction is not a negation of the divine but a necessary step in the unfolding of creation, where the infinite becomes manifest in the finite. However, this restriction also introduces the potential for fragmentation, as the divine light becomes increasingly concealed as it descends through the sefirot.

The material world, governed by the lowest of the sefirot, Malkhut, represents the densest and most fragmented level of emanation. Here, the divine light is so concealed that it can only be perceived indirectly, through the forms and structures of the physical world. Human beings, living in this material realm, often struggle to recognize the divine presence behind the veil of matter, leading to a sense of separation and disconnection from the Source. This fragmentation of light is the source of many spiritual challenges, as individuals must learn to see beyond the surface of material reality to reconnect with the higher emanations.

In Neoplatonism, Plotinus describes a similar process of descent, where the One emanates the Nous (Divine Intellect), which in turn gives rise to the Soul and the material world. As the emanations move further from the One, they lose their unity and become more divided. The material world, at the lowest level of this hierarchy, is characterized by multiplicity and impermanence, where the forms of existence are constantly changing and decaying. This constant flux creates a sense of instability and fragmentation, where the divine order is obscured by the chaotic nature of physical reality. Yet, even in this fragmented state, the material world retains a connection to the higher realms, as it is still an emanation of the divine.

The descent of emanations into the lower worlds also introduces the concept of limitation. In the higher realms, where the divine light flows freely, there are few boundaries or distinctions between beings. Consciousness is expansive, and the interconnectedness of all things is readily apparent. However, as emanations descend, they become subject to the limitations of space, time, and form. The material world, in particular, is governed by rigid structures and laws that impose boundaries on existence. These limitations create the conditions for individuality and differentiation, where each being and object appears separate and distinct from the others.

The experience of limitation in the material world is both a gift and a challenge. On one hand, it allows for the richness and diversity of creation, where each form of life can express its unique qualities and characteristics. On the other hand, these limitations often create a sense of isolation and separation, where individuals feel disconnected from the larger whole. The illusion of separateness, which arises from the fragmentation of divine light, is one of the greatest obstacles to spiritual awakening. It leads to the belief that the self is separate from the divine and from others, fostering desires, fears, and attachments that keep the soul bound to the material world.

In Gnosticism, this sense of limitation and fragmentation is understood as the fall of the soul into the material world, a realm created by the Demiurge. The Demiurge, unaware of the higher realms of the Pleroma, shapes the material world out of ignorance, creating a reality that is disconnected from the divine light. The soul, trapped in this world of illusion and limitation, forgets its origin in the divine and becomes entangled in the desires and sufferings of the flesh. Gnosticism teaches that the soul must awaken from this state of ignorance, recognizing the material world as a distortion of the true reality and seeking to return to the higher realms of light.

The transformation of emanations in the lower worlds also affects human consciousness. As divine light descends, consciousness becomes increasingly fragmented and identified with the material realm. In the higher spiritual realms, consciousness is expansive and unified, perceiving the interconnectedness of all things. However, in the material world, consciousness becomes bound by the ego, which creates a sense of separation between the self and the rest of existence. This fragmentation of consciousness is the root of many forms of suffering, as individuals feel isolated from both the divine and from others.

Yet, even in the midst of this fragmentation, the divine light is not completely lost. The material world, though dense and limited, is still an emanation of the divine, and the light of the Source can be found hidden within it. This idea is central to many mystical traditions, which teach that the task of spiritual practice is to reveal the hidden light within the material world and within the self. Through meditation, prayer, and acts of compassion, individuals can reconnect with the higher emanations, transforming their consciousness and the world around them.

One of the most profound effects of the transformation of emanations in the lower worlds is the creation of duality. As the divine light descends, it becomes divided into opposing forces— light and darkness, good and evil, spirit and matter. These dualities are not present in the higher realms, where the unity of the divine is absolute. However, in the material world, duality is a fundamental aspect of existence, shaping the way individuals perceive reality. The experience of duality creates the illusion of opposites, where the divine and the material, the spiritual and the physical, are seen as separate and often in conflict.

In many esoteric traditions, the resolution of this duality is one of the central goals of the spiritual path. The material world, with all its limitations and divisions, is not an obstacle to be rejected but a realm to be transformed. The divine light is present even in the darkest corners of existence, and the task of the soul is to unify the dualities of spirit and matter, revealing the underlying unity of all things. This process of unification is often described as spiritual alchemy, where the base elements of the material world are transmuted into the gold of divine consciousness.

The descent of emanations into the lower worlds also introduces the concept of multiplicity. In the higher realms, where the divine light is undivided, there is no distinction between beings or forms. All of existence is seen as a single, unified whole. However, as the emanations descend, they begin to fragment into individual forms and objects, creating the rich diversity of the material world. This multiplicity is a reflection of the infinite creativity of the divine, where each form of life expresses a different aspect of the divine mind.

However, this multiplicity also creates the potential for fragmentation and disconnection. In the material world, where beings are separate and distinct, the unity of the divine is often obscured. Individuals perceive themselves as isolated from one another and from the divine, leading to a sense of alienation and division. The spiritual path, therefore, involves seeing through the illusion of multiplicity to the underlying unity of all things. This realization allows individuals to recognize the divine presence in every form of existence and to experience the interconnectedness of all life.

The transformation of emanations in the lower worlds is, therefore, a complex and multifaceted process. As the divine light descends into the realms of matter, it becomes increasingly fragmented, limited, and concealed. Yet, even in this state of fragmentation, the divine is present, waiting to be revealed through acts of awareness, compassion, and spiritual practice. The task of the soul is to navigate the challenges of the material world, using the tools of meditation, contemplation, and service to uncover the hidden light within and to reunite with the higher emanations.

The descent of emanations into the lower worlds brings about significant changes, but these transformations are not final. Just as the divine light fragments and condenses to create the material realm, there exists the possibility of reversing this process, returning the scattered fragments of light to their original unity.

Spiritual traditions across the world emphasize the importance of practices that help reverse the densification and fragmentation of emanations. These practices can be understood as a kind of spiritual alchemy, where the lower, dense forms of consciousness are refined and elevated back toward the higher realms. The material world, with its limitations and multiplicity, becomes the stage upon which the soul works toward ascension, transforming itself and the world around it.

In alchemical traditions, this reversal of transformation is symbolized by the transmutation of base metals into gold. This process represents the purification and elevation of matter, turning what is dense and fragmented into something luminous and unified. However, alchemy is not just a physical process—it is also a metaphor for the inner work of the soul. The alchemist's quest to create the philosopher's stone mirrors the soul's journey to transform its lower nature and reconnect with the divine light. Through spiritual discipline, contemplation, and meditation, the soul works to dissolve the barriers of the ego and the illusions of separation, allowing the divine light within to shine forth.

In Kabbalah, this process is reflected in the concept of tikkun olam, or the "repair of the world." The idea of tikkun suggests that the universe, while fragmented, contains within it the potential for healing and reintegration. The divine sparks, scattered throughout creation during the process of emanation, can be gathered and restored through acts of awareness, kindness, and spiritual practice. Every action that brings light into the world—whether through ethical behavior, prayer, or the pursuit of wisdom—contributes to the repair of these divine fragments. In this way, the material world is not rejected but transformed, as the hidden light within it is revealed and elevated back toward the Source.

In Neoplatonism, Plotinus describes the ascent of the soul as a process of returning to the One, the ultimate source of all emanations. This ascent requires the soul to detach itself from the distractions and illusions of the material world, focusing instead on the eternal truths of the Nous (Divine Intellect). Through philosophical contemplation and inner purification, the soul gradually transcends the limitations of the physical realm and reconnects with the higher levels of reality. This process of ascent mirrors the descent of emanations, but it moves in the opposite direction, with the soul shedding the fragmentation of the lower worlds and merging back into the unity of the divine.

The concept of spiritual ascension is also central to the Gnostic tradition. In Gnosticism, the material world is seen as a prison, created by the Demiurge to entrap the soul in a state of ignorance and limitation. However, the soul contains within it a spark of the divine light, a remnant of its origin in the Pleroma. The goal of the Gnostic path is to awaken this divine spark through gnosis, a direct experience of divine knowledge. Once the soul recognizes its true nature, it can begin the process of ascension, escaping the illusions of the material world and returning to the higher realms of light. This ascension is not just an intellectual realization but a profound transformation, where the soul transcends the dualities of good and evil, light and darkness, and returns to a state of unity.

The reversal of the transformation of emanations is not just an individual journey—it is a cosmic process. In many esoteric traditions, the universe itself is seen as participating in the return to the Source. The process of emanation, while necessary for the creation of the world, also sets in motion the eventual return of all things to their original state of unity. This return is often described in terms of cycles, where the universe moves through phases of creation, dissolution, and reintegration. Time, in this sense, is not linear but cyclical, with each cycle bringing the universe closer to its ultimate reintegration with the divine.

In Kabbalistic thought, this cosmic process is reflected in the idea of the sefirot as both channels of emanation and paths of return. As the divine light descends through the sefirot, it becomes increasingly fragmented and concealed. However, through spiritual practice, the soul can ascend through the same sefirot, gradually reconnecting with the higher emanations. This ascent is not just a personal journey but part of the larger cosmic process of tikkun, where the entire universe is gradually healed and restored to its original unity with the Infinite (Ein Sof).

The concept of spiritual ascension also involves the idea of purification. As emanations descend, they become dense and fragmented, and this densification introduces impurities into the lower worlds. These impurities are not just material but also psychological and spiritual, manifesting as ignorance, attachment, and ego. The process of ascension, therefore, requires a purification of these impurities, where the soul gradually cleanses itself of the limitations and distortions introduced by the material world. This purification is often described as a "refining fire," where the soul passes through trials and challenges that burn away the dross of the lower nature, leaving only the pure light of the divine.

In alchemical traditions, this purification is symbolized by the process of calcination, where the base elements are subjected to intense heat, breaking down their impurities and revealing their true essence. The alchemist's work is not just to create physical gold but to transform the soul, refining it through the fire of spiritual discipline and contemplation. This inner alchemy mirrors the larger process of cosmic transformation, where the impurities of the material world are gradually purified and elevated back toward the divine light.

The role of consciousness is central to the process of reversing the transformation of emanations. In the lower worlds, consciousness becomes fragmented and identified with the ego, creating the illusion of separation from the divine. However, through spiritual awakening, consciousness can be expanded, allowing the individual to perceive the unity behind the multiplicity of the material world. This expansion of consciousness is the key to reversing the fragmentation of emanations, as it allows the individual to reconnect with the higher levels of reality and participate in the process of ascension.

In many mystical traditions, meditation is one of the primary tools for expanding consciousness and facilitating the reversal of transformation. Through meditation, the individual can quiet the distractions of the material world and enter into a state of inner stillness, where they can perceive the deeper realities of existence. This state of stillness allows the individual to transcend the limitations of the ego and experience the interconnectedness of all things. In this expanded state of consciousness, the divine light becomes more accessible, and the individual can begin the process of returning to the higher emanations.

The process of ascension is not limited to individual practice. Many traditions emphasize the importance of community and collective effort in reversing the transformation of emanations. In Kabbalah, for example, the concept of tikkun olam is not just about personal spiritual growth but about healing the world as a whole. Each act of kindness, each moment of spiritual awareness, contributes to the larger process of cosmic healing, helping to gather the scattered sparks of divine light and elevate them back toward the Source. In this way, the individual's spiritual journey is inseparable from the journey of the universe itself.

The reversal of transformation also involves the integration of opposites. In the lower worlds, dualities such as light and darkness, good and evil, spirit and matter, create the illusion of separation and conflict. However, the process of ascension involves the recognition that these dualities are not truly separate but are part of a larger unity. In alchemical terms, this is the reconciliation of opposites, where the soul learns to integrate the polarities of existence and transcend the limitations of dualistic thinking. This integration allows the soul to experience the unity of the divine, where all opposites are harmonized and dissolved into the oneness of the Source.

Ultimately, the transformation of emanations in the lower worlds is not a one-way descent but a dynamic process of descent and ascent, fragmentation and reintegration. While the descent of emanations creates the conditions for the material world, the reversal of this transformation allows for the healing and return of all things to the Source. This process of ascension is both a personal and cosmic journey, where individuals and the universe as a whole participate in the ongoing cycle of emanation and return. Through spiritual practice, inner work, and the expansion of consciousness, the divine light hidden within matter can be revealed and elevated, restoring the unity and purity of the original emanations.

Chapter 15
The Influence of Emanations on Daily Life

The theory of emanations, while often explored in philosophical and mystical contexts, extends beyond abstract metaphysics and enters the realm of daily life. The influence of emanations shapes not only the structure of the universe but also the experiences and actions of individuals within it.

At its core, the process of emanation describes the unfolding of divine energy from a central Source, gradually manifesting in increasingly complex and differentiated forms. This process does not stop at the level of abstract spiritual principles but extends all the way down to the physical world, influencing the patterns of nature and the rhythms of life. The seasons, the movement of celestial bodies, and the cycles of growth and decay are all reflections of the ebb and flow of emanations. The divine energy that originates in the Source permeates all aspects of existence, shaping both the macrocosmic and microcosmic dimensions of life.

In many esoteric traditions, the influence of emanations on daily life is seen in the natural world. The cycles of day and night, the phases of the moon, and the changing of the seasons are all considered to be expressions of divine energy moving through different levels of reality. These natural cycles mirror the larger process of emanation and return, where energy flows outward from the Source and then gradually returns to it. The rhythm of life is thus a reflection of the rhythm of the cosmos, where all things follow the patterns established by the emanations of divine energy.

Human beings, as part of the natural world, are deeply influenced by these cosmic patterns. The cycles of nature—birth, growth, decay, and renewal—are reflected in human life, both physically and spiritually. From a physical perspective, humans experience the passage of time, aging, and the cycles of sleep and wakefulness, all of which are shaped by the rhythms of nature. On a spiritual level, the soul moves through similar cycles of expansion and contraction, experiencing moments of clarity and illumination followed by periods of darkness and confusion. These fluctuations in consciousness are not random but are part of the larger process of emanation, where the soul participates in the flow of divine energy.

The influence of emanations on human emotions and thoughts is another key aspect of how this cosmic process manifests in daily life. Emotions, in many mystical traditions, are seen as currents of energy that reflect the movement of divine forces through the individual. Just as emanations descend through various levels of reality, emotions can range from the higher, more refined states of joy, love, and compassion to the lower, more fragmented states of fear, anger, and desire. These emotions are not separate from the spiritual process of emanation but are expressions of the same energy moving through different layers of the psyche.

In Kabbalah, the sefirot represent different aspects of divine energy that influence both the macrocosm of the universe and the microcosm of the individual soul. Each sefirah corresponds to a particular emotional or intellectual quality—such as love (Chesed), strength (Gevurah), or wisdom (Chokhmah)—and the balance between these energies shapes the internal life of the individual. When these energies are in harmony, the soul experiences peace, clarity, and connection to the divine. However, when they are out of balance, the soul may experience confusion, emotional turmoil, and a sense of disconnection from its true nature.

In this sense, the emanations that flow from the Source are not distant, abstract forces but are intimately connected to the inner life of each individual. The process of emanation is reflected in the ebb and flow of emotions, thoughts, and desires, all of which can either help or hinder the soul's journey of return.

By understanding how these emanations influence daily life, individuals can learn to navigate their inner landscape more effectively, aligning themselves with the higher emanations and avoiding the traps of lower, more fragmented energies.

One of the most practical applications of the theory of emanations in daily life is the idea of spiritual alignment. Many traditions teach that by aligning oneself with the higher emanations, individuals can experience greater clarity, peace, and connection to the divine. This alignment is achieved through practices such as meditation, prayer, and contemplation, which help to purify the mind and heart, allowing the individual to resonate with the higher frequencies of divine energy. When the soul is in alignment with these higher emanations, daily life becomes more harmonious, and the individual is better able to navigate challenges with grace and wisdom.

In Neoplatonism, Plotinus emphasizes the importance of philosophical contemplation as a means of aligning oneself with the higher levels of reality. Through contemplation, the individual can rise above the distractions of the sensory world and connect with the Nous (Divine Intellect), where the eternal truths of existence are revealed. This practice of contemplation is not just an intellectual exercise but a way of tuning the soul to the higher emanations, allowing the individual to experience a deeper sense of unity with the divine. As the soul aligns itself with the higher emanations, it becomes more attuned to the flow of divine energy in daily life, leading to greater clarity and insight.

Another way that emanations influence daily life is through the relationships between individuals. In many mystical traditions, relationships are seen as opportunities for the divine energy of emanations to be expressed and realized. The flow of love, compassion, and understanding between individuals is a reflection of the flow of divine energy between the higher and lower worlds. In this sense, relationships are not just human interactions but are spiritual exchanges, where the energy of the divine moves between people, helping them to grow and evolve.

However, relationships can also reflect the more fragmented aspects of emanations. When individuals are disconnected from the higher levels of reality, their relationships may become sources of conflict, misunderstanding, and pain. These negative dynamics often arise when individuals are operating from a place of ego and separation, rather than from a place of unity and connection to the divine. By recognizing the influence of emanations on relationships, individuals can learn to cultivate more harmonious interactions, aligning their relationships with the flow of divine energy and using them as opportunities for spiritual growth.

The influence of emanations on decision-making is another important aspect of how this cosmic process manifests in daily life. Every decision an individual makes is influenced by the interplay of divine energy and the limitations of the material world. When individuals are aligned with the higher emanations, their decisions are guided by wisdom, compassion, and a sense of connection to the larger whole. However, when individuals are disconnected from these higher levels, their decisions may be driven by fear, desire, or ego, leading to outcomes that create more fragmentation and suffering.

In esoteric traditions, the process of decision-making is often seen as an opportunity to align oneself with the divine will. By tuning into the higher emanations and listening to the inner voice of intuition, individuals can make decisions that are in harmony with the flow of divine energy. This practice of spiritual discernment helps individuals to navigate the complexities of daily life with greater clarity and purpose, allowing them to make choices that support their spiritual evolution and contribute to the larger process of cosmic healing.

In many ways, the influence of emanations on daily life can be understood as a two-way process. Just as emanations flow downward from the Source, shaping the structure of the universe and the experiences of individuals, human actions and decisions can also influence the flow of emanations. When individuals act with awareness, compassion, and a sense of connection to the divine, they contribute to the elevation of divine energy in the world. These actions help to gather the scattered fragments of light and return them to the Source, participating in the larger process of healing and reintegration.

Ultimately, the influence of emanations on daily life reveals the interconnectedness of all things. Every thought, emotion, and action is part of the larger flow of divine energy, moving through the cycles of creation, dissolution, and return. By becoming aware of how emanations shape their experiences, individuals can learn to align themselves with the higher levels of reality, using their daily lives as a path of spiritual growth and transformation. The material world, rather than being a distraction from the spiritual path, becomes a mirror of the divine, reflecting the flow of emanations in every moment.

The influence of emanations on daily life goes beyond understanding the cycles of nature or human emotions—it permeates every facet of existence. When individuals begin to recognize the presence of these cosmic forces at work, they can learn to live in a way that harmonizes with the flow of emanations, transforming mundane experiences into moments of spiritual insight.

One of the most powerful ways to harness the influence of emanations in daily life is through the practice of mindfulness. By being fully present in each moment, individuals can become more aware of the subtle energies that flow through their thoughts, emotions, and actions. Mindfulness, in this context, is not just about calming the mind or reducing stress; it is about tuning into the larger cosmic currents that influence human experience. When practiced with the awareness of emanations, mindfulness becomes a tool for aligning oneself with the higher flows of energy that originate from the Source, allowing individuals to act with greater clarity and purpose.

In many spiritual traditions, mindfulness is linked to the concept of presence—the state of being fully conscious and engaged with the present moment, free from the distractions of past regrets or future anxieties. In the context of the Theory of

Emanations, presence allows individuals to perceive the divine energy that permeates every aspect of their surroundings. When the mind is still and open, the flow of emanations becomes more evident, revealing the interconnectedness of all things. This awareness transforms daily tasks, making them opportunities for spiritual reflection and growth. Simple activities, such as walking, eating, or conversing with others, become sacred when approached with an understanding of their connection to the divine emanations that sustain the universe.

Meditation, contemplation, and prayer also play a crucial role in strengthening one's connection to the flow of emanations in daily life. These practices help clear the mind of distractions, allowing the individual to tap into the higher emanations that flow from the Source. In meditation, one can move beyond the surface level of thoughts and emotions, entering into a deeper awareness of the divine light within and around them. This expanded state of consciousness helps the individual become a more receptive vessel for the higher emanations, which can then influence their actions, decisions, and relationships.

In Kabbalistic thought, these practices are often seen as tools for ascending the sefirot and aligning oneself with the higher aspects of divine energy. Prayer, in particular, is viewed as a way of reaching out to the divine, not as a petition for material gains but as a means of harmonizing one's will with the divine will. When approached in this way, prayer becomes a dialogue with the higher emanations, helping the individual align their consciousness with the flow of divine light. Similarly, contemplation of sacred texts or spiritual teachings allows the mind to reflect on eternal truths, elevating the soul's awareness beyond the immediate concerns of the material world.

The practice of gratitude is another way to connect with the flow of emanations in daily life. Gratitude shifts the focus from lack or dissatisfaction to the abundance of divine energy that constantly flows through existence. By recognizing and giving thanks for the blessings, both large and small, that emanate from the Source, individuals open themselves up to receiving more of this divine energy. Gratitude also fosters a sense of humility, reminding the individual that they are part of a larger cosmic process and that the emanations that sustain their life come from a higher power. This attitude of humility and gratitude strengthens the connection between the individual and the divine, making them more attuned to the subtle currents of energy that shape their experiences.

Another key aspect of living in harmony with emanations is the practice of conscious action. Every action an individual takes has the potential to either align with the flow of divine energy or create more fragmentation and separation. Conscious action means acting with awareness of the higher purpose behind one's decisions, seeking to reflect the divine qualities of love, wisdom, and compassion in every interaction. This practice is especially important in moments of conflict or difficulty, where the tendency may be to act out of fear, anger, or ego. By pausing and reflecting on how their actions can serve the greater good, individuals can align themselves with the higher emanations and contribute to the cosmic process of healing and reintegration.

In many esoteric traditions, conscious action is linked to the concept of karma, where every action generates a corresponding effect that reverberates through the fabric of reality. By acting in alignment with the higher emanations, individuals generate positive karma, contributing to the elevation of divine energy in the world. Conversely, actions driven by ego, fear, or ignorance create negative karma, leading to further fragmentation and suffering. The knowledge of emanations helps individuals understand that their actions are not isolated but are part of a larger web of cause and effect, where every choice has the potential to either elevate or diminish the flow of divine energy in the world.

Relationships also serve as a powerful reflection of the influence of emanations in daily life. Interactions with others, whether harmonious or conflictual, mirror the dynamics of cosmic energy. When individuals approach relationships with a sense of unity and compassion, they allow the higher emanations to flow freely between themselves and others, creating bonds that are nourishing and supportive. However, when relationships are based on ego, control, or fear, they reflect the lower, more fragmented levels of emanation, leading to disconnection and suffering.

In Kabbalistic tradition, relationships are seen as opportunities for tikkun—the healing and restoration of divine light in the world. Every act of kindness, every moment of understanding, is a chance to repair the fragmentation introduced by the descent of emanations into the material world. By cultivating empathy, compassion, and forgiveness in relationships, individuals can contribute to the process of cosmic healing, helping to gather the scattered sparks of divine light and elevate them back toward the Source.

The influence of emanations is also evident in the pursuit of creative endeavors. Art, music, literature, and other forms of creative expression can be seen as channels through which divine energy flows into the world. When individuals create from a place of inspiration, they are tapping into the higher emanations, bringing forth beauty, truth, and harmony that reflect the divine order. Creative acts that resonate with the higher realms have the power to elevate both the creator and the audience, providing glimpses of the unity and transcendence that lie beyond the material world. In this way, creativity becomes a form of spiritual practice, where the act of creation mirrors the larger process of cosmic emanation and return.

Finally, the influence of emanations can be applied to the way individuals interact with the natural world. Nature, as an expression of divine emanation, reflects the flow of energy from the Source into the physical realm. By cultivating a sense of reverence and respect for the natural world, individuals can deepen their connection to the higher emanations that flow through all of creation. This awareness can manifest in simple acts of environmental stewardship—such as caring for the earth, protecting wildlife, or living sustainably—which align with the flow of divine energy and contribute to the preservation of the balance and harmony of nature.

Ultimately, the influence of emanations on daily life reveals the profound interconnectedness of all things. The divine energy that flows from the Source shapes not only the cosmos but also the experiences, actions, and relationships of individuals in the material world. By cultivating awareness of this flow and aligning themselves with the higher emanations, individuals can transform their daily lives into a spiritual practice, where every moment becomes an opportunity to reflect the divine light and contribute to the healing and reintegration of the universe.

Chapter 16
Unity in Emanations

The concept of unity is central to understanding the nature of emanations. Even as the divine energy flows outward from the Source and becomes increasingly diverse, fragmented, and differentiated, there remains an underlying unity that binds all levels of reality. This unity is not easily perceived in the lower worlds, where separation and multiplicity seem to dominate, but it is present nonetheless—woven into the very fabric of existence.

At the highest level, where the emanations originate directly from the Source, unity is absolute. In many mystical traditions, this primordial unity is described as the One, the Infinite, or the Unmanifest. It is beyond all dualities and distinctions, a state of perfect oneness where all things exist in a unified, undivided form. In Neoplatonism, this is the realm of the One, from which all emanations flow. The One is beyond time, space, and form—yet it is the cause of all that exists. From this perspective, unity is not something that is achieved or created but is the natural state of being at the highest level of reality.

As emanations descend from the Source, the unity of the One becomes progressively differentiated. In the realm of the Divine Intellect, or Nous, the first distinctions begin to emerge. This level of reality contains the archetypal patterns of all things, yet these patterns still exist in a state of harmony and interconnection. While distinctions are present, they do not disrupt the underlying unity. This is the first step in the unfolding of the cosmos, where the unity of the divine begins to express itself in the form of multiplicity without losing its essential harmony.

In Kabbalistic thought, this process is represented by the sefirot, the ten attributes or emanations through which the Infinite manifests in the world. Each sefirah represents a distinct quality—such as wisdom, understanding, or mercy—but all are interconnected and flow from the same divine Source. The sefirot are often depicted as a tree, with each branch and root representing a different aspect of the divine unity. While each sefirah has its own unique function, they all work together as part of a larger whole, reflecting the unity that underlies the multiplicity of existence.

As the emanations continue to descend, they enter the realm of the material world, where the unity of the divine becomes increasingly concealed. In the physical realm, multiplicity and separation seem to dominate. Each form of life appears distinct and separate from others, and the interconnectedness of all things is often obscured by the limitations of time, space, and matter. The divine unity is still present, but it is hidden beneath the surface of material reality, waiting to be revealed through spiritual practice and inner reflection.

In human experience, this underlying unity is often felt as a deep sense of connection to the world and to others. Moments of spiritual insight, profound love, or creative inspiration can reveal glimpses of this unity, where the boundaries between the self and the world dissolve. These experiences are often described as moments of oneness, where the individual feels connected to something greater than themselves. This sense of unity is a reflection of the divine light that flows through all things, reminding the soul of its origin in the Source.

In the context of relationships, the idea of unity plays a crucial role. While individuals may appear separate and distinct, the flow of emanations connects them on a deeper level. Relationships are often described as mirrors of the divine, where the qualities of love, compassion, and understanding reflect the underlying unity of all beings. When approached with this awareness, relationships become opportunities to experience the interconnectedness of life, transcending the illusion of separation and fostering a deeper sense of unity with others.

In many spiritual traditions, the path to awakening involves recognizing and embracing this unity. The process of spiritual growth is often described as a return to oneness, where the soul gradually overcomes the illusions of separation and reconnects with the divine Source. This return to unity is not about rejecting the material world or denying the reality of multiplicity but about seeing through the apparent divisions to the deeper unity that underlies all things.

In Sufism, this journey toward unity is expressed through the concept of tawhid, the recognition of the oneness of God. The Sufi path involves dissolving the ego, or the false sense of separation, and merging with the divine presence. This process of unification is not an abstract philosophical idea but a deeply personal and experiential journey, where the Sufi seeks to experience the divine in every moment and in every aspect of life. Through practices such as dhikr (remembrance of God) and muraqaba (meditation), the Sufi aligns themselves with the flow of divine emanations, gradually dissolving the boundaries between the self and the divine.

In a similar way, the mystics of many other traditions—whether Christian, Hindu, or Buddhist—speak of the journey toward unity as the ultimate goal of the spiritual path. The realization of this unity is often described as a state of enlightenment or liberation, where the individual no longer identifies with the separate self but recognizes their true nature as part of the divine whole. This recognition is not just intellectual but a transformation of consciousness, where the soul directly experiences the unity that permeates all things.

While the perception of unity is often obscured in the lower worlds, it is never entirely lost. Even in the densest and most fragmented levels of existence, the divine light remains present. The task of the soul is to awaken to this hidden unity, to see beyond the surface of reality and recognize the interconnectedness of all life. This awakening is both a personal and a collective process, as each soul's realization of unity contributes to the larger cosmic return to the Source.

The unity in emanations also has practical implications for how individuals live their lives. When one recognizes the interconnectedness of all things, it changes the way they relate to the world. Actions are no longer seen as isolated events but as part of a larger web of cause and effect, where each choice contributes to the greater whole. This awareness encourages a sense of responsibility, compassion, and empathy, as individuals begin to understand that their actions affect not only themselves but the entire cosmos.

Living in alignment with the unity of emanations involves cultivating practices that foster connection and harmony, both within oneself and with the world. Meditation, prayer, acts of service, and moments of reflection help individuals stay attuned to the flow of divine energy, allowing them to act in ways that support the healing and reintegration of the fragmented aspects of existence. Each action, when done with awareness of the underlying unity, becomes an opportunity to participate in the cosmic process of return.

Thus, while the diversity and multiplicity of the material world can create the illusion of separation, the deeper reality is one of unity. This unity is not only a metaphysical truth but a lived experience, one that can be realized through spiritual practice, mindful living, and a deepening awareness of the divine presence that flows through all things.

The profound unity that underlies all emanations, though often obscured in the material world, is a central theme across mystical traditions. As emanations descend from the Source and multiply into various forms, they may appear increasingly distinct. However, beneath this seeming multiplicity, the essential connection to the One remains unbroken. This unity, though veiled by the complexities of material existence, can be rediscovered through spiritual practices that cultivate awareness of the interconnectedness of all things.

A key challenge in perceiving the unity of emanations lies in the layers of separation that appear as the divine energy moves further from its Source. In the physical world, distinctions between individuals, objects, and experiences can easily overshadow the subtle bonds that unite them. Yet, as explored earlier, all things are expressions of the same divine energy. The task of recognizing this unity, then, is about shifting one's perception—seeing beyond the immediate appearances of separation and realizing that all beings share the same origin and are part of the same cosmic flow.

One path toward recognizing this unity is through practices that still the mind and draw attention away from surface-level distractions. Meditation, in particular, offers a method for transcending the fragmentation of daily life and experiencing the underlying harmony of all things. By focusing the mind on a single point—such as the breath, a sacred word, or an image of divine light—meditation creates space for the deeper currents of divine energy to emerge. In this state of inner stillness, the divisions between self and world begin to dissolve, and the unity of emanations becomes more apparent.

In traditions such as Buddhism, this dissolution of boundaries is reflected in the concept of emptiness or sunyata. Emptiness does not imply nothingness but rather the absence of inherent separation. All things are interconnected, and their distinctness is illusory—like waves on the surface of the ocean, which are all part of the same vast body of water. Through meditation, one comes to recognize this truth, realizing that the distinctions between self and others, between the physical and the spiritual, are not fixed. This insight into the emptiness of phenomena helps the practitioner to experience the unity that flows through all of existence.

Another practice that fosters awareness of unity is the cultivation of compassion. In many spiritual traditions, compassion is seen as a reflection of the divine love that permeates the cosmos. By practicing compassion toward others, one affirms the interconnectedness of all beings, breaking down the barriers of ego and separation. Acts of kindness and empathy are not merely ethical behaviors but expressions of the recognition that all life is connected. Each compassionate act strengthens the awareness of unity, reminding the individual that the boundaries between self and other are ultimately illusory.

In Kabbalistic thought, the practice of chesed (loving-kindness) reflects this principle of unity. As one of the sefirot, chesed represents the boundless love and mercy of the divine, which flows through all levels of creation. When individuals act with loving-kindness, they align themselves with this divine energy, helping to bridge the gaps created by the descent of emanations. Every act of compassion becomes an opportunity to restore the unity that has been obscured by the fragmentation of the material world.

Contemplative practices that focus on the interconnectedness of nature also provide a way to reconnect with the unity of emanations. Nature, as an expression of divine energy, reveals the intricate web of life in which all things are interdependent. By observing the cycles of nature—such as the turning of the seasons, the growth of plants, or the flow of rivers—individuals can come to appreciate the subtle rhythms that mirror the process of emanation and return. This awareness fosters a deeper sense of connection to the world, revealing the unity that lies beneath the surface of all natural forms.

In many indigenous traditions, the connection between humanity and nature is seen as a direct reflection of the unity that permeates the cosmos. Rituals honoring the earth, the elements, and the cycles of life are ways of recognizing that all beings are part of the same sacred whole. This recognition encourages a sense of stewardship and reverence for the natural world, as every aspect of creation is understood to be infused with divine energy. By living in harmony with the earth, individuals not only sustain the physical environment but also align themselves with the flow of emanations, participating in the larger process of cosmic balance.

Another dimension of realizing unity in daily life involves the practice of non-attachment. Many spiritual traditions teach that attachment to material forms—whether to objects, people, or ideas—creates a sense of separation, reinforcing the illusion of division between the self and the world. By cultivating non-attachment, individuals learn to see beyond the surface of material reality and recognize the deeper unity that binds all things together. This does not mean rejecting the material world but rather approaching it with a sense of openness and fluidity, understanding that all forms are temporary manifestations of the same underlying energy.

In Hinduism, the concept of maya describes the illusion that arises from attachment to the physical world. Maya creates the sense that the material realm is separate from the divine, leading to ignorance of the true nature of reality. The spiritual path, then, involves piercing through the veil of maya to experience the unity of the divine. Practices such as yoga and meditation are tools for transcending the illusions of separation and reconnecting with the oneness that lies beneath the surface of existence.

Unity in emanations also has a profound impact on how individuals experience time. In the lower worlds, time often appears linear, with events following one another in a sequence of past, present, and future. However, at higher levels of emanation, time is experienced as cyclical or even timeless, where all moments exist simultaneously in the eternal present. This shift in perspective allows individuals to transcend the limitations of chronological time and experience life as part of a larger, interconnected flow of energy. By embracing this sense of timelessness, individuals can release anxieties about the future or regrets about the past, living fully in the present moment where the unity of emanations is most directly felt.

In mystical traditions, the concept of the "eternal now" reflects this timeless aspect of unity. When one is fully present, free from the constraints of linear time, the boundaries between moments dissolve, and the unity of all experiences becomes apparent. Practices such as mindfulness and deep contemplation help cultivate this awareness of timelessness, allowing individuals to touch the eternal aspects of their being and experience the divine unity that permeates all things.

The realization of unity in emanations also invites individuals to approach their actions with greater awareness of their cosmic significance. Every action, thought, and intention contributes to the larger flow of divine energy, influencing not only the individual but also the entire web of existence. By acting with mindfulness and intention, individuals can align themselves with the higher emanations, contributing to the healing and restoration of the fragmented aspects of the cosmos. This awareness transforms daily life into a sacred practice, where every moment offers the opportunity to reflect the unity of the divine.

In sum, the unity in emanations is not something that exists only at the highest levels of reality—it is a truth that permeates all layers of existence, from the Source down to the material world. Though often obscured by the multiplicity and fragmentation of the physical realm, this unity can be rediscovered through practices that foster awareness, compassion, and non-attachment. By embracing the interconnectedness of all things, individuals can live in greater harmony with the flow of emanations, experiencing the divine unity that sustains all of creation.

Chapter 17
Emanations and Freedom

The relationship between emanations and freedom is intricate and layered. As the divine energy flows outward from the Source and manifests in the lower worlds, questions arise about the role of freedom—both cosmic and individual—in this process. The more distant emanations become from the Source, the more they seem bound by the constraints of form, time, and matter. Yet, within this limitation, there exists a profound potential for freedom.

At the highest level of emanation, in the realm closest to the Source, freedom is understood as absolute. The Source itself is infinite, boundless, and unconstrained by any external forces. It is the origin of all things, yet it exists beyond any need for movement, change, or division. In this state, freedom is not the result of choice or opposition but the natural state of being. There is no duality, no restriction—only pure, limitless existence. The freedom of the Source is complete because it is undivided and all-encompassing. This divine freedom reflects the ultimate potential of all emanations.

As the emanations flow outward, passing through the layers of reality, freedom begins to take on new dimensions. In the higher spiritual realms, where divine intellect and will are at work, freedom is still vast, though it begins to operate within the framework of cosmic order and purpose. The Divine Intellect (Nous) directs the flow of emanations, shaping the structure of reality with wisdom and clarity. In this realm, freedom is expressed through the harmonious unfolding of divine will, where all beings are aligned with the greater cosmic plan.

At this level, freedom is not understood as the ability to act against divine will but as the ability to express one's true nature in perfect harmony with the Source. This is the freedom of alignment—where beings act in accordance with their highest purpose, free from the distortions of ego or fear. In Neoplatonic thought, this is the realm where the soul contemplates eternal truths and experiences the fullness of freedom by aligning itself with the divine order.

As emanations continue to descend into the material world, the nature of freedom becomes more complex. In the lower worlds, where duality and fragmentation dominate, freedom often appears to be limited by external constraints—time, space, physical laws, and the demands of the body. In this realm, freedom is no longer absolute but is conditioned by the limitations of the material plane. Human beings, in particular, experience freedom as the capacity for choice—often framed in terms of moral, philosophical, or existential decisions. This form of freedom, though profound, is often constrained by the conditions of material existence.

In Kabbalistic thought, this tension between freedom and limitation is reflected in the relationship between the sefirot of Chesed (loving-kindness) and Gevurah (strength, judgment). Chesed represents expansive, unlimited freedom—the divine force of generosity and openness. Gevurah, on the other hand, represents boundaries, discipline, and restriction. The balance between these two sefirot is essential for understanding how freedom operates in the world. While Chesed allows for boundless expression, Gevurah provides the necessary structure that gives form and direction to that expression. Together, they create the conditions for true freedom—freedom that is not chaotic or destructive but purposeful and aligned with divine wisdom.

In the material world, the experience of freedom is often tied to the concept of free will—the ability to make choices that shape one's life and destiny. Yet, from the perspective of emanations, free will is not simply about choosing between good and evil or following one's desires. It is about aligning one's choices with the flow of divine energy, recognizing the interconnectedness of all actions and their cosmic significance. True freedom, in this sense, is the ability to act in harmony with the higher emanations, participating consciously in the unfolding of the divine plan.

However, the material world, with its fragmentation and density, also introduces the possibility of disconnection from the Source. In this realm, the soul may experience a sense of separation and isolation, leading to choices driven by ego, fear, or desire. These choices, while seemingly expressions of freedom, often lead to further bondage—binding the soul to the illusions of the material world. This paradox of freedom is central to the spiritual journey: while the soul has the capacity for free will, the choices it makes can either bring it closer to true freedom or deepen its sense of limitation.

In many mystical traditions, the path to spiritual freedom involves recognizing the illusions of the material world and reconnecting with the higher emanations. This process of awakening is often described as a return to the original state of unity, where the soul once again experiences the boundless freedom of the Source. This journey requires the dissolution of the ego, the transcendence of duality, and the cultivation of awareness that sees beyond the surface of material existence. As the soul aligns itself with the higher emanations, it regains its connection to the infinite freedom that flows from the divine.

In Gnosticism, this return to freedom is framed as the liberation of the soul from the prison of the material world, created by the Demiurge. The Gnostic path involves the acquisition of gnosis, or divine knowledge, which allows the soul to recognize its true nature and escape the limitations of the physical realm. Freedom, in this context, is not just the ability to choose but the ability to see through the illusions of the lower worlds and reconnect with the divine Source.

In a similar way, the yogic traditions of India describe freedom (moksha) as the release from the cycles of birth, death, and rebirth (samsara). This liberation is achieved through self-realization, where the individual soul (atman) recognizes its oneness with the ultimate reality (Brahman). The freedom of moksha is not about acting according to personal desires but about transcending the egoic self and experiencing the infinite, unconditioned freedom of the divine.

In the material world, the experience of freedom is also influenced by the laws of karma—the cosmic principle of cause and effect. Every action creates a ripple in the fabric of reality, shaping the conditions of future experiences. In this sense, freedom is not only the capacity to act but also the responsibility to understand the consequences of one's actions. When individuals act in alignment with the higher emanations, their actions contribute to the healing and reintegration of the fragmented aspects of the cosmos. However, when actions are driven by ego or ignorance, they create further fragmentation, binding the soul to the cycles of karma.

The understanding of freedom in relation to emanations also invites a reconsideration of the nature of power and control. In the lower worlds, power is often equated with the ability to dominate or control others, shaping events according to one's will. Yet, from the perspective of emanations, true power comes from alignment with the divine flow. It is not about forcing outcomes but about attuning oneself to the natural rhythms of the cosmos, acting with wisdom and compassion. This form of freedom is not about exerting control over external circumstances but about cultivating inner freedom—freedom from fear, desire, and the illusion of separateness.

In practical terms, this understanding of freedom invites individuals to live with greater awareness of their choices and their impact on the world. Every decision becomes an opportunity to align with the higher emanations, to act in harmony with the divine flow, and to contribute to the healing and restoration of the cosmic order. This awareness transforms the experience of freedom from a burden of choice into a path of conscious participation in the divine plan.

Thus, the journey toward freedom within the framework of emanations is not about escaping the limitations of the material world but about transcending the illusions that bind the soul. True freedom is found in the recognition of unity, in the alignment of will with the divine, and in the conscious participation in the unfolding of creation. The soul, through its choices, has the potential to reflect the boundless freedom of the Source even within the constraints of the material plane.

As the exploration of freedom within the framework of emanations deepens, the tension between free will and divine will becomes more apparent.

In many traditions, the material world is often seen as a testing ground for the soul's capacity to exercise free will. The soul, inhabiting a body and navigating a world of form, faces numerous distractions and influences that can pull it away from its original connection to the divine. The choices made in this realm often reflect either a movement toward or away from this connection. The question of freedom becomes not merely about external circumstances but about the internal alignment with the higher forces of emanation.

One way to understand this dynamic is to consider the difference between superficial freedom and deep spiritual freedom. Superficial freedom, as it is often experienced in the material world, involves the ability to make choices based on desires, preferences, and external conditions. This kind of freedom is bound by the limitations of the ego, which perceives itself as separate from the world around it. While this freedom appears expansive, it is often constrained by attachment to outcomes, fear of loss, and a sense of lack. The ego, operating from a fragmented understanding of reality, seeks control over circumstances and results, but this desire for control ultimately leads to further entanglement in the material plane.

In contrast, deep spiritual freedom involves a surrender of the ego's need for control and an embrace of the natural flow of divine energy. This form of freedom is not about avoiding choices but about making choices from a place of alignment with the higher emanations. The individual who attains this level of freedom no longer views their actions as separate from the greater cosmic order but recognizes that every decision and action is part of a larger, interconnected whole. Acting from this awareness allows the individual to move beyond fear, attachment, and desire, embracing the freedom that comes from unity with the divine will.

In the context of emanations, freedom is closely tied to the process of returning to the Source. As the soul moves through the layers of emanation, it experiences varying degrees of limitation and separation. The lower the soul descends into the material world, the more fragmented its perception of reality becomes. However, the soul retains within it the potential to reconnect with the higher emanations and, ultimately, with the Source itself. This reconnection is the essence of spiritual freedom—the realization that the soul is not bound by the limitations of the material world but is capable of transcending them through awareness and alignment.

Many mystical traditions emphasize the importance of self-knowledge in attaining this level of freedom. In Gnosticism, for example, the path to liberation lies in acquiring knowledge (gnosis) of one's divine origin. This knowledge allows the soul to see through the illusions of the material world and recognize its true nature as part of the divine. The soul, having descended into the lower realms of emanation, becomes entangled in the physical plane, but through gnosis, it can awaken to its higher purpose and begin the journey of return. Freedom, in this sense, is the ability to see beyond the surface of material existence and to act in accordance with the deeper truth of the soul's connection to the divine.

In Kabbalistic teachings, the concept of teshuvah—often translated as repentance or return—reflects this process of spiritual freedom. Teshuvah is not merely about correcting mistakes but about returning to the original state of unity with the divine. It involves a conscious turning away from the distractions and distortions of the material world and a return to the higher emanations from which the soul originates. This return is not a one-time event but a continual process of realignment, where the soul repeatedly seeks to reconnect with the divine light despite the challenges of the physical realm. Each act of teshuvah strengthens the soul's sense of freedom, as it moves closer to its true, unbounded nature.

A key element of this process is the dissolution of the ego. The ego, as the construct of the lower self, operates from a perspective of separation and limitation. It identifies with the body, the mind, and the external circumstances of life, creating a sense of individuality that is often at odds with the unity of the divine. The ego's desire for control, power, and recognition keeps the soul bound to the lower realms of emanation, where freedom is experienced as a constant struggle against external forces. However, as the soul begins to dissolve the ego through spiritual practice, it becomes more attuned to the higher emanations and experiences a deeper sense of freedom—freedom not as control but as surrender to the divine flow.

This surrender is not passive but requires a heightened level of awareness and discernment. Acting in alignment with the divine will does not mean relinquishing all personal agency but rather refining one's choices to reflect the greater cosmic order. In this sense, true freedom involves a balance between individual will and divine will, where the individual's actions become expressions of the divine energy flowing through them. This is the freedom of co-creation, where the soul participates consciously in the unfolding of the cosmos, guided by the wisdom of the higher emanations.

One way to cultivate this level of freedom is through the practice of inner stillness. Meditation, contemplation, and prayer all provide opportunities for the soul to quiet the distractions of the ego and attune to the subtle currents of divine energy. In moments of stillness, the boundaries between the self and the world begin to dissolve, allowing the individual to experience the deeper unity that underlies all things. This state of inner stillness is where true freedom is found, as the individual is no longer bound by the fluctuations of thought, emotion, or circumstance but is rooted in the eternal presence of the divine.

In many spiritual traditions, this experience of stillness is described as an encounter with the divine light. The light, as an expression of the highest emanations, illuminates the soul's path, revealing both its limitations and its potential for freedom. The more the soul becomes receptive to this light, the more it is able to transcend the illusions of the material world and experience the freedom that comes from unity with the Source. This encounter with the divine light is transformative, as it not only brings clarity and insight but also empowers the soul to act with greater wisdom and compassion.

The concept of freedom in the context of emanations also invites a reconsideration of the role of fate and destiny. In the material world, individuals often feel bound by external circumstances, believing that their lives are shaped by forces beyond their control. However, from the perspective of emanations, fate is not a rigid, predetermined path but a dynamic interplay between divine will and individual choice. The soul, though influenced by cosmic forces, retains the ability to shape its own destiny through conscious alignment with the higher emanations. This understanding of freedom transforms the experience of life from one of passive acceptance to one of active participation in the divine plan.

Ultimately, the journey toward spiritual freedom is not about escaping the material world but about transforming one's relationship to it. The material realm, though dense and fragmented, is still an expression of divine energy. By recognizing this and aligning one's actions with the flow of emanations, the soul can experience freedom even within the constraints of the physical plane. This freedom is not about avoiding challenges or difficulties but about approaching them with a deeper awareness of their role in the soul's evolution. Each moment, each choice, becomes an opportunity to reflect the divine light and move closer to the ultimate freedom of unity with the Source.

Chapter 18
The Redemption of Fragmented Emanations

The idea of fragmentation within the process of emanation introduces a profound mystery: how can something that originates from the One, the Source of all unity, become scattered and divided? As divine energy moves outward from the Source into the lower realms, it undergoes a series of transformations, gradually becoming denser and more complex. In the material world, this process results in the appearance of multiplicity, separation, and even fragmentation. However, this fragmentation is not an ultimate reality but a temporary state—a necessary stage in the cosmic journey of descent and return. Redemption, then, is the process of gathering these scattered pieces, restoring the unity that underlies the apparent division.

The concept of fragmented emanations is particularly emphasized in Kabbalistic thought, where it is understood through the idea of the breaking of the vessels (Shevirat HaKelim). According to this mystical teaching, the divine light, in its purest form, was too intense for the vessels designed to contain it. As a result, the vessels shattered, and the sparks of divine light were scattered throughout creation, becoming embedded in the material world. These sparks, though concealed, retain their connection to the Source, and it is the task of humanity to elevate them, bringing about their redemption.

The process of redemption, in this context, is not simply about spiritual salvation on an individual level but involves a cosmic reintegration. Every action, every thought, every moment is an opportunity to gather these divine sparks and return them to their origin. This concept transforms the understanding of daily life, as it suggests that even in the most mundane activities, there exists the potential for profound spiritual work. The fragmentation of divine energy is not a mistake or a flaw in creation, but part of a larger process that invites conscious participation in the healing and reintegration of the universe.

In the material world, where fragmentation is most apparent, the soul often feels the effects of separation. The sense of being divided—from others, from the world, and even from oneself—is a common human experience. This feeling of disconnection reflects the fragmentation of emanations, where the unity of the Source is obscured by the multiplicity of forms. The journey of the soul, therefore, involves recognizing these fragments and working to restore their unity. This process is both personal and collective, as the redemption of individual souls contributes to the larger cosmic return.

One of the primary ways to engage in the work of redemption is through the cultivation of awareness. The more the soul becomes aware of its true nature as a manifestation of the divine, the more it can recognize the divine presence in the world around it. This awareness allows the soul to see beyond the surface-level fragmentation and perceive the hidden unity that connects all things. By cultivating this awareness, individuals can begin to act in ways that reflect the deeper truth of unity, rather than the illusions of separation.

In spiritual traditions such as Sufism, the process of redemption is often described as the journey from multiplicity to unity. The Sufi path involves dissolving the ego, or the false sense of separation, and returning to a state of oneness with the divine. This journey is not linear but cyclical, as the soul moves through stages of expansion and contraction, continually deepening its understanding of unity. Through practices such as dhikr (remembrance of God) and tawhid (the affirmation of divine oneness), the Sufi seeks to gather the fragmented aspects of their being and restore them to their original unity.

In a similar way, many mystical traditions speak of redemption as the process of returning to wholeness. In the Christian tradition, the idea of salvation involves the reconciliation of the individual soul with God, restoring the relationship that was broken by sin or separation. This process is not just about forgiveness but about healing the divisions within the soul, allowing it to once again reflect the unity of the divine. The work of redemption is seen as both a personal and collective task, as the salvation of each individual contributes to the restoration of the entire creation.

The concept of spiritual alchemy also reflects this idea of redemption. In alchemical traditions, the process of transforming base metals into gold is a metaphor for the transformation of the soul. The fragmented, lower aspects of the self—represented by the base metals—are purified and elevated through spiritual work, ultimately returning to their original state of unity with the divine. This process of transmutation involves both the dissolution of the ego and the reintegration of the fragmented aspects of the self, allowing the soul to reflect the light of the higher emanations.

The redemption of fragmented emanations is not only a spiritual or metaphysical concept but also has practical implications for how individuals live their lives. Every interaction, every relationship, every choice has the potential to either contribute to or hinder the process of cosmic reintegration. When individuals act from a place of love, compassion, and awareness, they help to gather the scattered sparks of divine light and return them to the Source. Conversely, actions driven by fear, hatred, or ignorance reinforce the fragmentation, deepening the sense of separation.

In this sense, redemption is not a passive process but an active engagement with the world. The work of gathering the sparks requires mindfulness and intentionality, as individuals seek to align their actions with the flow of divine energy. This work can take many forms—acts of kindness, moments of deep contemplation, creative expression, or simply being present in the world with an open heart. Each of these actions contributes to the healing of the fragmented aspects of creation, bringing them closer to their original state of unity.

One of the central teachings of this process is that redemption is available in every moment. The divine sparks are hidden in the most ordinary aspects of life, waiting to be discovered and elevated. This teaching encourages a shift in perspective, where the material world is no longer seen as a distraction from spiritual work but as the very arena in which that work takes place. The world, with all its challenges and imperfections, becomes the stage upon which the drama of cosmic redemption unfolds.

However, the process of redemption is not without its challenges. The material world, with its dense and fragmented nature, often obscures the presence of the divine. The ego, with its attachment to separation and control, resists the work of reintegration. Yet, it is precisely within these challenges that the potential for redemption lies. The difficulties of life are not obstacles to spiritual growth but opportunities to engage in the work of gathering the scattered sparks and restoring them to their original unity.

In many mystical traditions, this work of redemption is understood as part of a larger cosmic cycle. The descent of the emanations into the material world is not a one-way process but part of a dynamic flow that includes both descent and return. The fragmentation that occurs in the lower worlds is not permanent but part of the unfolding journey of creation. The work of redemption is the work of helping the universe complete this cycle, bringing all things back into alignment with the divine order.

Thus, the redemption of fragmented emanations involves both a personal and cosmic dimension. On a personal level, it is the process of healing the divisions within the self, recognizing the unity that underlies all aspects of being. On a cosmic level, it is the work of participating in the restoration of the entire creation, helping to gather the scattered pieces of divine light and return them to the Source. This work is both a privilege and a responsibility, as every soul plays a part in the unfolding story of cosmic redemption.

The work of redeeming fragmented emanations reaches deeper than a mere spiritual journey for the individual soul. It touches the entire cosmos, weaving together the scattered pieces of divine light and slowly restoring the original unity that was once present. In this process, spiritual practices and mystical traditions offer specific ways to engage with the brokenness of the material world, allowing individuals to consciously participate in the cosmic healing.

In Kabbalistic tradition, the notion of tikkun olam, or the repair of the world, captures the essence of the redemptive process. Every action, thought, or intention has the power to affect not just the individual soul but the fabric of creation itself. Each small act of kindness, each moment of insight, gathers and uplifts the divine sparks embedded in material reality. The individual, through their spiritual and moral choices, becomes an active participant in the grand work of tikkun. This process is continuous—there is always more light to uncover, more fragments to heal, as the soul navigates its earthly existence.

The work of tikkun emphasizes the profound interconnectedness of all beings and all actions. Redemption is not isolated to personal salvation or individual enlightenment; it is a collective effort that spans across time and space. Every person, every event, every moment is part of a larger cosmic network, and each contributes to the healing of the whole. This interconnectedness is a reflection of the original unity that exists at the highest levels of emanation, before the descent into multiplicity and fragmentation. In this sense, every act of healing—whether through meditation, prayer, or compassionate service—ripples outward, contributing to the greater restoration of the universe.

Spiritual traditions around the world have long recognized the importance of healing, both on a personal and cosmic level. In Christian mysticism, the process of redemption is understood as the restoration of the relationship between humanity and God, fractured by sin and separation. The path to redemption is through Christ, whose sacrifice represents the ultimate act of reconciliation, mending the rift between the divine and the material. This redemptive act is seen as universal, applying to all of creation, and it invites believers to participate in the ongoing process of spiritual healing by embodying the qualities of love, forgiveness, and humility.

In Sufism, the notion of fana—the annihilation of the self in the presence of God—represents a key aspect of redemption. The fragmented ego, with its attachment to worldly desires and separateness, is dissolved in the divine presence. Through this dissolution, the soul experiences union with the One, and in this unity, it contributes to the healing of the cosmos. The Sufi practices of dhikr (remembrance) and sama (listening) are designed to bring the soul into alignment with the divine, stripping away the layers of ego and illusion that obscure the true unity underlying all existence. Through this alignment, the Sufi becomes a vessel for divine light, participating in the work of cosmic redemption.

The practices of spiritual alchemy also provide insight into the redemptive process. In alchemical traditions, the transformation of base metals into gold serves as a metaphor for the spiritual purification and elevation of the soul. The fragmented, lower aspects of the self—represented by the base metals—must undergo a series of transmutations, being purified and refined until they are restored to their original, divine state. This process mirrors the larger cosmic work of gathering the scattered divine sparks and returning them to the Source. The alchemist's journey is not just one of personal enlightenment but of contributing to the greater healing of the universe.

In Hinduism, the concept of moksha—liberation from the cycle of birth, death, and rebirth—reflects a similar theme of redemption. The soul, bound by the material world and the illusions of separateness, must awaken to its true nature and return to its original unity with Brahman, the ultimate reality. The practices of yoga, meditation, and devotion are paths that lead to this realization, helping the soul to transcend the fragmentation of earthly existence and experience the wholeness of divine union. Through the realization of moksha, the soul is not only freed from its individual karmic bonds but also contributes to the cosmic balance, participating in the larger process of return.

In this grand work of redemption, there is a recognition that the world itself—dense, material, and often filled with suffering—is the very place where the divine light is concealed and where the work of healing must be done. The mystics and sages of many traditions understand that the path to spiritual enlightenment is not about escaping the world but about transforming one's relationship to it. The material world, with all its challenges and complexities, is the stage upon which the soul's journey of redemption unfolds. Each obstacle, each moment of suffering, is an opportunity to uncover the hidden divine light and return it to its source.

The work of redemption, therefore, requires both patience and perseverance. The fragmentation of the divine light occurred over vast stretches of time, and its reintegration is an ongoing, gradual process. The soul must engage in this work with humility, recognizing that redemption is not achieved through grand, dramatic gestures but through the steady, consistent practice of love, compassion, and mindfulness. Each small act of healing—whether through a kind word, a moment of reflection, or an offering of service—contributes to the larger cosmic work of reintegration.

In the context of the material world, this process of redemption also invites individuals to cultivate a deeper awareness of their surroundings. The divine sparks are not hidden in some distant, abstract realm; they are present in the most ordinary aspects of life. The challenge is to see beyond the surface-level fragmentation and recognize the divine presence that permeates all things. This awareness transforms the way individuals interact with the world, encouraging them to approach every situation with a sense of reverence and purpose.

By cultivating this awareness, individuals can begin to engage with the world in a way that reflects the unity of all things. This shift in perspective allows for a more compassionate, empathetic approach to life, as the boundaries between self and other, between the sacred and the mundane, begin to dissolve. In this way, the work of redemption becomes not just a spiritual practice but a way of being in the world—one that honors the divine light in every person, every creature, every experience.

The practices that support this work of redemption are varied, but they all share a common goal: to reconnect the soul with the higher emanations and, through this reconnection, contribute to the healing of the cosmos. Meditation, prayer, ritual, and acts of service are all tools that help individuals align with the flow of divine energy, gathering the scattered sparks and returning them to the Source. Each of these practices, when performed with intention and awareness, becomes a sacred act of participation in the larger cosmic drama of redemption.

Thus, the redemption of fragmented emanations is not simply a personal journey of spiritual growth but a collective process that involves all of creation. The divine light, though scattered and concealed, remains present in every aspect of the world, waiting to be uncovered and uplifted. Through conscious engagement with the world, individuals can participate in the grand work of redemption, helping to restore the unity that underlies all things. This work, though challenging, is also deeply fulfilling, as it allows the soul to experience its true nature as a reflection of the divine and to contribute to the healing and reintegration of the cosmos.

Chapter 19
The Future of Emanations

As the cosmic journey of emanation unfolds, a question lingers on the horizon: What lies in the future of these divine emanations? The process of emanation, descent into multiplicity, and eventual return to the Source is not static—it is a dynamic movement, a cosmic cycle that continually evolves.

In many esoteric teachings, the future of emanations is seen as a return to the state of pure unity from which they originated. However, this return is not merely a reversal of the initial emanation. The movement toward unity involves a transformation and elevation of the entire cosmos. While the universe begins in simplicity and unity, the descent into multiplicity and complexity creates a richness of experience that is not lost in the return to the One. The journey through fragmentation and materiality, with all its trials, brings with it the potential for a deeper understanding and a more profound unity— one that incorporates the diversity of creation.

In Neoplatonic thought, the ultimate return of emanations to the Source is described as the henosis, or the mystical union of all beings with the One. The soul, having descended into the material world and experienced the fragmentation of existence, seeks to ascend back through the layers of emanation, purifying itself along the way. This ascent leads to a final merging with the divine, where all distinctions dissolve into the infinite unity of the Source. Yet, this return is not a dissolution into nothingness; it is a return to fullness, where the soul, now enriched by its journey, participates in the divine in a more complete and conscious way.

In this vision, the future of emanations is not about the annihilation of individuality but about the transcendence of separateness. The diversity of the created world is not erased but integrated into a higher unity. Each individual soul, having experienced the uniqueness of its own path, contributes to the larger cosmic harmony. The return to the Source, then, is a reconciliation of opposites—a bringing together of the many into the One, where the distinctions of form and substance are transcended, yet the essence of each remains.

In Kabbalistic tradition, the future of emanations is understood through the concept of Olam HaBa, the World to Come. This eschatological vision describes a time when the divine light that was scattered throughout creation will be fully restored to its original unity. The breaking of the vessels, which caused the fragmentation of the divine light, will be repaired through the process of tikkun olam—the repair of the world. In this future state, the divine sparks that have been elevated through human action and spiritual practice will return to their source, and the world itself will be transformed into a place of divine harmony and peace.

The World to Come is often described as a time of enlightenment, where the barriers between the spiritual and the material are lifted, and the divine presence becomes fully manifest in the world. In this future state, the hidden aspects of the divine that were once obscured by the density of the material world are revealed, and the relationship between creation and the Creator is restored to its original purity. The future of emanations, in this sense, is the culmination of the process of redemption, where all of creation is brought back into alignment with the divine will.

The mystical traditions of Islam, particularly Sufism, also contemplate the future of emanations in terms of the soul's return to God. The Sufi path involves a series of stages, where the soul moves through increasing levels of purification and enlightenment. The ultimate goal of this path is the realization of fana—the annihilation of the ego and the merging of the soul with the divine presence. In the final stages of this journey, the soul experiences baqa—the state of eternal subsistence in God. Here, the individual no longer perceives themselves as separate from the divine but as a reflection of the One. This vision of the future involves not only the return of the individual soul to God but the transformation of all creation as it is absorbed back into the divine reality.

In Hindu cosmology, the future of emanations is framed by the concept of cosmic cycles, or yugas. The universe is understood to pass through a series of ages, each representing a different phase in the cosmic process. As the universe progresses through these cycles, it moves from a state of purity and enlightenment to increasing levels of fragmentation and darkness. However, this descent is not permanent. At the end of each cycle, the universe undergoes a process of dissolution (pralaya), after which it is renewed and begins a new cycle of creation. This cosmic rhythm reflects the ongoing process of emanation and return, where the universe continually moves through cycles of unity and fragmentation, each time returning to the Source before emerging anew.

The idea of cosmic cycles also appears in Buddhist thought, where the process of samsara—the cycle of birth, death, and rebirth—represents the continual movement of souls through various levels of existence. The ultimate goal in Buddhism is to transcend samsara and achieve nirvana—a state of liberation from the cycles of suffering and ignorance. In the context of emanations, nirvana can be understood as the return of the soul to its original state of unity, where all attachments to the material world are dissolved, and the soul experiences the bliss of enlightenment.

However, the future of emanations is not only about the soul's individual journey. Many spiritual traditions also speak of a collective transformation, where the entire cosmos is elevated and returned to its original state of harmony. This collective return is often described in eschatological terms, as a time when the divine presence becomes fully manifest, and the world is restored to its intended perfection. In Christianity, this vision is reflected in the concept of the Kingdom of God, where the world is transformed through divine grace, and all creation is reconciled with God. This future state is one of peace, justice, and unity, where the separation between heaven and earth is overcome.

In this sense, the future of emanations is not just a return to the past but a movement toward a higher realization of unity. The experiences of fragmentation, separation, and diversity are not erased but integrated into a new, more profound understanding of oneness. The universe, having passed through the stages of emanation, descent, and redemption, moves toward a state of fulfillment, where the divine plan is fully realized.

What makes this future of emanations particularly compelling is the notion that the journey itself is meaningful. The descent into multiplicity, with all its challenges and trials, serves a purpose in the larger cosmic process. Through the experiences of separation and return, the soul—and indeed the entire cosmos—gains a deeper understanding of the divine. The return to unity is not a simple return to the beginning, but a return enriched by the journey, where the fullness of creation is realized within the oneness of the Source.

The eschatological visions of the future of emanations, then, invite individuals to participate in the unfolding of this cosmic drama. The future is not fixed or predetermined but is shaped by the choices and actions of souls as they navigate the complexities of existence. Each act of redemption, each moment of awareness, contributes to the larger process of return, helping to bring about the final restoration of all things to the divine unity.

Thus, the future of emanations is a story of hope and transformation. It speaks of a time when the fragmentation and suffering of the material world will be healed, and the divine light will shine through all of creation once again. This vision calls individuals to engage with the world in a way that reflects the future they hope to see—one of unity, peace, and divine harmony.

The journey of emanations through time leads us into the final stage of their cosmic cycle: the return to the ultimate unity from which all things emerged. In this final phase, the idea of universal reintegration emerges as a profound reflection on the fate of all creation. This process envisions a future in which the myriad expressions of the Source, having descended through countless layers of reality, reverse their path, gradually dissolving back into the original unity. However, this reintegration does not mean an erasure of individuality but an elevation of all creation into a harmonious state of being, where diversity and unity coexist in perfect balance.

In mystical traditions, this notion of final unity is more than a theoretical concept—it is seen as the culmination of a cosmic drama that has been unfolding since the first moment of emanation. The future of the universe, therefore, is not merely a return to an earlier state of being but a new form of existence, transformed and enriched by the experiences gained through descent into multiplicity. This enriched unity, sometimes referred to as the consummation of all things, suggests that everything in the universe has a purpose, and that purpose is fulfilled in the final return to the Source.

In esoteric interpretations, the concept of universal reintegration often parallels the idea of cosmic cycles that appear in many traditions. These cycles describe not only the periodic creation and destruction of worlds but also the ongoing renewal of spiritual reality. Each cosmic cycle—whether understood as a grand epoch or a spiritual era—represents a movement away from the Source and a subsequent return. As the cycles progress, the souls and beings that emerge from the emanations experience growth, fragmentation, and ultimately, reunification. The future of emanations, then, is bound to these cycles, and the final stage represents the closing of one grand cycle before another may begin.

In Neoplatonism, the return to unity, or henosis, is described as the soul's movement from the fragmented world of matter back through the higher realms of intellect and spirit, until it reenters the undifferentiated unity of the One. The entire cosmos participates in this ascent, and the future of emanations is envisioned as a universal return, where all beings, both spiritual and material, are reconciled with their divine origin. This reconciliation involves a process of purification and elevation, as the material elements of creation are transformed and absorbed into higher states of existence.

In Kabbalah, this process is captured in the idea of tikkun olam, the repair and restoration of the world. The final stage of this process is not merely the healing of individual souls but the reintegration of all the divine sparks scattered throughout creation. When these sparks are gathered and returned to their source, the world itself undergoes a transformation, becoming a vessel for divine light. This final state of harmony and peace is described as a time when the hidden aspects of the divine are fully revealed, and the material world is aligned with spiritual reality. In this vision, the future of emanations leads to the complete manifestation of divine presence in all things, where the veil between the spiritual and material is lifted, and creation returns to a state of divine transparency.

In this eschatological vision, time and space, as we know them, cease to be constraints on the experience of unity. The fragmentation that characterizes existence in the material world is transcended, and all beings experience their oneness with the Source. The temporal and spatial distinctions that define life in the lower realms dissolve, and what remains is the eternal presence of the divine, where past, present, and future merge into a single, all-encompassing reality.

Mystical traditions often describe this final state as one of profound peace, joy, and bliss—a state where the soul, having completed its journey through the layers of existence, rests in the eternal presence of the divine. This resting, however, is not a static condition but a dynamic and ongoing experience of union with the Source. The future of emanations, in this sense, is a continual unfolding of divine reality, where the soul participates in the infinite expression of the divine will. Each moment of this union is a reflection of the eternal, and the soul, now free from the limitations of matter, experiences the fullness of divine love, wisdom, and power.

This final unity is not limited to the individual soul or even to humanity. It encompasses all levels of existence, from the highest spiritual beings to the material elements of creation. In this grand vision, the entire cosmos is seen as a reflection of the divine, and the future of emanations is the restoration of that reflection to its original clarity. Every part of creation, from the smallest particle to the grandest star, is involved in this process, and all things return to the Source, bringing with them the richness and diversity of their journey through time and space.

The return to unity also brings with it a new understanding of freedom. As beings ascend through the levels of emanation, their experience of freedom becomes more profound. In the lower realms, freedom is often limited by the constraints of matter, time, and circumstance. However, as the soul moves closer to the Source, it experiences a freedom that is not bound by external conditions but is rooted in the very essence of divine being. This ultimate freedom is the freedom to fully express one's divine nature, to act in perfect alignment with the will of the Source, and to participate in the ongoing creation and renewal of the universe.

In this vision of the future, even the concept of individuality is transformed. The soul does not lose its identity in the return to the Source but transcends the egoic boundaries that once defined it. Individuality becomes an expression of the divine in its infinite diversity, and each soul, while united with the One, retains its unique qualities as a reflection of the divine. This individuality, however, is no longer experienced as separation but as a harmonious part of the greater whole, where all beings exist in a state of mutual interdependence and unity.

The future of emanations, therefore, is a story of return, transformation, and fulfillment. It is the culmination of the cosmic process that began with the first moment of emanation and will continue to unfold as all things move toward their ultimate reunion with the Source. This return is not the end but a new beginning, where creation, having completed its cycle of descent and return, enters into a new phase of existence—one that is marked by the full revelation of the divine in all things.

Chapter 20
Practices for Connecting with Higher Emanations

The intricate journey of emanations, descending from the Source and diffusing into multiplicity, creates a vast spectrum of realities, each with its own level of spiritual density. As beings move through these levels, the question naturally arises: how can individuals reconnect with the higher emanations that reflect the purest expressions of divine light?

Mystical traditions across the world offer various practices that serve as bridges to the higher realms. One of the most universal methods is meditation. Through meditation, the mind is gradually quieted, allowing the soul to attune to higher frequencies of reality that are otherwise overshadowed by the noise of everyday life. By turning inward and focusing on the inner light, the practitioner begins to shift their consciousness from the external, fragmented world toward the unified presence of the higher emanations.

In some esoteric traditions, meditation is not merely a passive state but an active process of alignment. The practitioner intentionally focuses their awareness on specific divine attributes or emanations, such as wisdom, love, or will, attempting to draw these qualities into their own being. This practice, known as contemplative absorption, brings the soul into resonance with the higher emanations, as the individual becomes a vessel through which divine light flows. The deeper the meditation, the more the boundaries between the individual and the higher realms dissolve, allowing the practitioner to experience a direct connection with the emanating energies.

Another key practice is prayer, which serves as both a petition and a means of communion with the higher realms. While prayer often involves the act of asking for guidance, protection, or assistance, it also provides a framework for the soul to reach upward and connect with the divine source of emanations. In many traditions, the act of prayer is not only about seeking divine intervention but about aligning one's own will with the will of the Source. This alignment creates a direct channel through which higher emanations can flow, allowing the soul to receive the light and wisdom of the higher planes.

In Kabbalistic tradition, specific meditations and prayers, often based on sacred texts such as the Zohar, are designed to elevate the practitioner's consciousness through the sefirot, the ten emanations of divine attributes. Each sefirah represents a different aspect of divine light, and by meditating on these, the practitioner seeks to ascend through the layers of spiritual reality, moving closer to the ultimate unity of the Source. This practice is a disciplined path, requiring both focus and a deep understanding of the spiritual structures that underlie creation. The goal is not simply to reach the higher emanations but to integrate their light into everyday life, transforming the individual into a living reflection of the divine.

Visualization is another powerful tool for connecting with higher emanations. In many mystical traditions, visualization practices involve the mental creation of images or symbols that represent the divine presence or higher spiritual realities. These images serve as focal points for the mind, helping to lift the practitioner's awareness out of the material plane and into the realm of the spiritual. The act of visualizing light, divine beings, or sacred geometries can activate latent spiritual faculties within the soul, creating a direct link to the higher emanations that these images represent.

In Tibetan Buddhism, for instance, the practice of deity yoga involves visualizing oneself as a divine being, embodying the qualities and attributes of that deity. This process of identification serves to dissolve the egoic sense of separation, allowing the practitioner to connect more deeply with the divine emanations that flow through the universe. By visualizing oneself as a conduit for divine light, the practitioner opens the way for higher energies to infuse their being, leading to a greater sense of unity with the Source.

Breathing exercises are also used to align the soul with higher emanations. In many spiritual traditions, breath is seen as a bridge between the physical and spiritual worlds. The act of conscious breathing—sometimes referred to as pranayama in yogic traditions—serves to purify the mind and body, preparing the practitioner to receive the finer energies of the higher planes. By controlling the breath, the practitioner can still the mind, calm the emotions, and open the spiritual channels through which divine light flows. Each breath becomes an act of communion with the Source, as the inhalation draws in divine energy and the exhalation releases the impurities of the material world.

Ritual and ceremony also play a significant role in connecting with higher emanations. Throughout history, human beings have used sacred rites as a way to align themselves with divine forces. These rituals often involve symbolic actions, offerings, chants, and invocations, all of which serve to open a channel between the practitioner and the higher realms. In many traditions, rituals are carefully structured to mirror the patterns of the cosmos, creating a resonance between the microcosm of the ritual space and the macrocosm of the divine universe. By participating in these sacred ceremonies, the practitioner places themselves in harmony with the larger cosmic order, facilitating the flow of divine energy into the material world.

One such example can be found in the rituals of ancient Egyptian religion, where ceremonies designed to invoke the presence of divine beings were a central part of spiritual practice. These rituals were not only about honoring the gods but about creating an environment where the divine could manifest on earth. The use of sacred words, symbols, and gestures allowed the practitioners to align themselves with the higher emanations that flowed from the gods, bringing those divine forces into their own lives and into the world around them.

Another example is found in Christian mysticism, where the Eucharist is seen as a means of direct communion with the divine. The act of partaking in the body and blood of Christ is not merely a symbolic gesture but is understood as a way of integrating the divine presence into one's own being. Through this sacred ritual, the practitioner connects with the higher emanations of love, grace, and salvation, experiencing a profound unity with the divine.

In addition to structured practices like meditation and prayer, there are also more spontaneous methods of connecting with higher emanations. These often involve moments of deep contemplation, where the individual simply rests in the presence of the divine, allowing their awareness to expand and merge with the higher realms. This practice requires an openness and receptivity to the flow of divine energy, without the need for formal structures or techniques. In these moments, the soul surrenders to the higher emanations, allowing them to penetrate and transform the deeper layers of consciousness.

Ultimately, the key to connecting with higher emanations lies in cultivating a state of inner alignment with the divine. This alignment is not a one-time achievement but an ongoing process, as the soul continually refines its ability to resonate with the higher planes. Through meditation, prayer, visualization, and ritual, the practitioner creates pathways for divine light to enter their being, lifting them out of the fragmentation of the material world and into the harmonious flow of the emanations.

These practices, when approached with dedication and intention, serve to dissolve the barriers that separate the soul from its divine origin. As the soul becomes more attuned to the higher emanations, it begins to reflect those emanations in its thoughts, actions, and relationships. Each practice becomes a step in the ascent toward unity with the Source, bringing the practitioner closer to the ultimate realization of their true nature as a reflection of the divine.

As the exploration of practices for connecting with higher emanations continues, the focus now shifts to advanced methods—those that require deeper spiritual maturity and a heightened sensitivity to the subtle energies of the higher planes. These practices are designed to not only foster connection but to sustain and integrate the experience of higher emanations into daily life, allowing individuals to live in alignment with the divine flow.

One of the most profound methods for maintaining this connection is the practice of spiritual contemplation. In contrast to active meditation, where the mind often focuses on a particular symbol, word, or concept, contemplation involves resting the mind in silence and stillness, allowing it to be fully receptive to the emanations of the higher planes. In this state of quiet surrender, the individual becomes an open vessel for divine energy, without the need to direct or control the experience. The practice of contemplation invites the practitioner to move beyond thought, beyond the intellect, and into a direct experience of the divine presence.

This method is often described in the writings of Christian mystics such as John of the Cross and Teresa of Ávila, who refer to it as the "prayer of quiet" or "infused contemplation." Here, the soul is drawn into the presence of God not by effort but by grace, experiencing a profound union with the divine that is both peaceful and transformative. The soul, in this state, is fully attuned to the higher emanations, experiencing a sense of being held in the divine embrace. This practice allows the individual to remain in constant awareness of the Source, even in the midst of daily activities.

Another advanced practice is the cultivation of constant remembrance. In Sufi mysticism, this is known as dhikr, the practice of continuously remembering and invoking the presence of the divine throughout all aspects of life. This remembrance can take the form of silent repetition of divine names, sacred phrases, or simply the inward turning of attention toward the divine presence in every moment. The goal of dhikr is not only to create moments of divine connection but to transform one's entire existence into an unbroken flow of divine awareness.

In this practice, the boundaries between the sacred and the mundane begin to dissolve. The practitioner learns to see the divine in every experience, every interaction, every breath. By continually focusing on the higher emanations, the soul remains aligned with the flow of divine energy, no longer pulled into the distractions and fragmentation of the material world. Dhikr becomes a way of living in harmony with the higher planes, allowing the soul to reflect the light of the Source in every aspect of its being.

Another powerful practice for sustaining the connection with higher emanations is the visualization of divine light. This technique is often used in mystical traditions to bring the light of the higher realms into the practitioner's body and mind. In advanced stages, this practice involves not only visualizing divine light but embodying it, becoming a conduit through which the higher emanations can flow into the world. The practitioner visualizes their entire being as filled with radiant light, dissolving any barriers between the self and the divine. This practice is particularly potent for those who seek to integrate their spiritual experiences into their physical and emotional lives.

In Kabbalistic mysticism, such visualizations often focus on the sefirot, where each emanation is seen as a distinct form of divine light. The practitioner visualizes these lights descending from the highest realms, passing through each sefirah and filling their body with divine energy. This process not only connects the individual with the higher emanations but also purifies and elevates their consciousness, allowing them to align more deeply with the divine flow.

For those who have established a foundation of spiritual practices, self-transcendence becomes a central goal. The more the practitioner immerses themselves in the higher emanations, the less they identify with the limitations of the ego or the material world. This does not mean a rejection of the material realm but rather a transcendence of its illusions. In this state, the practitioner no longer experiences themselves as a separate entity but as an expression of the divine energy that flows through all things.

The state of self-transcendence is described in various mystical traditions, where it is often seen as the pinnacle of spiritual practice. In Hinduism, it is reflected in the concept of moksha—liberation from the cycle of birth and rebirth. In this state, the soul has fully realized its identity with the divine and is no longer bound by the constraints of the material world. Similarly, in Buddhist teachings, this state is known as nirvana, where the illusions of the self and the material world are extinguished, and the soul experiences complete unity with the Source.

In the process of transcending the self, the practitioner also experiences a deepening of their connection with all of creation. The realization that all beings are manifestations of the same divine light leads to a profound sense of compassion and empathy. This awareness transforms the way the practitioner interacts with the world, as they begin to see all actions, relationships, and experiences as opportunities to express and reflect the higher emanations. Compassionate action becomes a natural expression of the soul's alignment with the divine, as the practitioner seeks to elevate not only their own consciousness but the consciousness of those around them.

For many advanced practitioners, service becomes an integral part of their spiritual path. The experience of connecting with higher emanations leads to a desire to share that divine energy with the world, to act as a channel through which the light of the higher planes can flow into the material realm. In this way, spiritual practice is no longer confined to meditation or prayer but extends into every act of service, every moment of care for others. This is the ultimate expression of spiritual integration—the ability to bring the light of the higher emanations into every aspect of life, transforming the world through the power of divine love.

The mystical path, as it ascends through these advanced practices, ultimately leads to the dissolution of all perceived separation. The practitioner who has sustained their connection with the higher emanations gradually comes to realize that they are not simply reaching upward toward a distant divine source, but that they themselves are a part of that Source. The boundaries between self and the divine, between the higher planes and the material world, between subject and object, begin to blur and dissolve. What remains is an all-encompassing unity, where the practitioner exists in perfect harmony with the flow of emanations.

These advanced practices, while powerful, also require great care and discernment. As the soul moves closer to the higher emanations, it must remain vigilant against the subtle traps of spiritual pride or attachment to mystical experiences. The goal of these practices is not to achieve special states or powers but to align more deeply with the divine flow, to become a pure reflection of the Source. Humility, patience, and a continual surrender to the divine will are essential qualities on this path, as the practitioner learns to release their own desires and ambitions in favor of the greater cosmic purpose.

Ultimately, the practices of connecting with higher emanations are not ends in themselves but pathways toward a deeper realization of unity with the Source. Through meditation, contemplation, prayer, visualization, and service, the soul gradually ascends through the layers of emanation, shedding the illusions of separation and returning to its original state of divine harmony. Each step on this journey brings the soul closer to the ultimate goal—a state of being where the individual reflects the pure light of the Source, living in perfect alignment with the higher emanations, and participating in the unfolding cosmic dance of creation and return.

Chapter 21
Enlightenment through Emanations

The concept of enlightenment is deeply intertwined with the journey of emanations. As the soul travels through the layers of existence, descending from the Source and engaging with the complexities of the material world, it seeks not only to return but to do so with a fuller understanding of the divine truth. Enlightenment, in this context, represents the soul's awakening to its true nature as an emanation of the divine, and its conscious alignment with the flow of higher emanations.

The path to enlightenment is often depicted as a gradual process of awakening. The soul, bound by the illusions of the material world and the fragmentation of reality, begins its journey in a state of spiritual sleep. At this level, the soul identifies primarily with its physical body and the external world, unaware of its deeper connection to the higher realms. This stage, often called ignorance or avidya in some traditions, marks the beginning of the soul's journey, where its divine origin is obscured by the density of the material plane.

The first step toward enlightenment involves what can be described as the initial awakening—a moment of realization that there is more to existence than the material world. This stage is often triggered by a spiritual experience, a deep insight, or an encounter with divine energy that shakes the soul out of its complacency. The individual begins to perceive the world differently, sensing that behind the multiplicity of forms and experiences, there is a deeper, unifying reality. This awakening may come suddenly, or it may emerge gradually, as the soul becomes more attuned to the presence of the higher emanations that flow through all things.

Once the soul has tasted this initial awakening, it enters into a period of spiritual seeking. At this stage, the soul is filled with a desire to understand the nature of reality, to connect more deeply with the higher planes, and to experience the divine directly. This seeking can take many forms—study, meditation, prayer, ritual—but at its core, it is driven by the soul's innate longing to return to its Source. The seeker begins to explore various spiritual paths and teachings, learning to quiet the mind and open the heart to the subtle energies of the higher realms.

As the soul progresses on this path, it begins to experience moments of illumination—glimpses of the divine light that permeates all creation. These moments may occur during meditation, in nature, or in the midst of everyday life, when the veils of illusion briefly lift and the soul perceives the unity that underlies all things. These illuminations serve as guideposts on the path to enlightenment, reminding the soul of its true nature and encouraging it to continue its ascent. However, they are fleeting, and the soul must learn not to cling to these experiences but to use them as stepping stones toward a deeper, more sustained realization.

One of the central themes in the path to enlightenment is the purification of the soul. As the individual moves closer to the higher emanations, they must shed the impurities of ego, desire, and attachment that bind them to the lower realms. This purification is often described as a process of inner alchemy, where the soul is refined and transformed, much like base metal being transmuted into gold. The practices of meditation, contemplation, and service are all tools for this purification, as they help the soul dissolve the barriers that separate it from the divine.

At this stage, the soul may experience periods of intense struggle, often referred to as the dark night of the soul. In this phase, the individual feels distant from the divine, as if all the spiritual progress they have made has been lost. This experience, while painful, is an essential part of the journey, as it forces the soul to confront its deepest fears and attachments. Through this process of purification, the soul is gradually stripped of all that is not in alignment with the higher emanations, emerging from the darkness with a deeper understanding of its own divine nature.

Once the soul has passed through this purification, it begins to experience sustained illumination. At this level, the individual no longer experiences divine light as fleeting moments of insight but as a constant presence that permeates their consciousness. The soul is now firmly aligned with the flow of higher emanations, and the barriers between the self and the divine begin to dissolve. The individual sees the divine in all things and experiences a deep sense of unity with the Source. This stage is often referred to as the enlightened state—a state of being where the soul is fully awake to its true nature and lives in harmony with the divine flow.

However, even this stage is not the final goal. In many mystical traditions, enlightenment is seen not as an endpoint but as the beginning of a new phase of spiritual development. The enlightened soul is now called to serve the world, using its connection to the higher emanations to uplift others and bring healing to the material plane. This is the stage of the bodhisattva in Buddhist tradition, where the enlightened being chooses to remain in the world to help all sentient beings awaken to their true nature. In this phase, the individual becomes a living embodiment of the divine, radiating light, love, and wisdom to all those they encounter.

In the context of the Theory of Emanations, enlightenment represents the soul's conscious return to its origin, but with a new understanding of its place within the cosmic order. The soul, having descended through the layers of emanation and experienced the fragmentation of the material world, now ascends back to the Source, carrying with it the wisdom gained through its journey. This ascent is not merely a reversal of the descent but an elevation, where the soul transcends the limitations of duality and multiplicity, experiencing the unity of the divine in a more profound and complete way.

The enlightened soul no longer perceives itself as separate from the Source but as an expression of the divine emanations that flow through all things. This realization brings with it a sense of freedom, as the soul is no longer bound by the illusions of time, space, and ego. It moves through the world with a deep sense of peace and joy, knowing that it is part of the greater cosmic dance of emanation and return.

In this state of enlightenment, the soul experiences a deep connection with the higher emanations, not as something external to itself but as its own true nature. The boundaries between self and other, between the material and the spiritual, dissolve, and the soul rests in the knowledge that it is one with the divine. This is the ultimate goal of the journey of emanations—to realize the unity that has always existed beneath the surface of multiplicity and to live in full alignment with the divine will.

The journey is unique for each individual, and progress is often marked by moments of breakthrough followed by periods of consolidation and integration. The path to enlightenment requires patience, dedication, and a willingness to surrender to the flow of the higher emanations. Yet, for those who persist, the reward is the realization of their own divine nature and the experience of living in harmony with the Source.

As the soul moves through these stages, it begins to understand that enlightenment is not a distant goal to be achieved but a state of being that is available in every moment. By continually aligning with the higher emanations, the individual can experience the peace and joy of enlightenment even in the midst of the challenges of the material world. In this way, the journey of emanations becomes a path not only of return but of transformation, where the soul, through its awakening, brings the light of the divine into every aspect of existence.

The culmination of the journey through the emanations leads the soul toward a state of deep, transformative enlightenment—an awakening that is not merely an individual achievement but a realization that transcends personal identity and merges with the cosmic order. In this final chapter, we explore the profound nature of enlightenment as the ultimate union with the Absolute Principle, drawing upon the experiences of spiritual masters and enlightened beings who have embodied this state.

Enlightenment, in its fullest sense, is the moment when the soul not only perceives the divine reality but becomes a living reflection of it. The barriers that once separated the individual from the Source—ego, desire, and the fragmentation of the material world—are dissolved. The soul stands in the light of pure awareness, recognizing itself as one with the flow of emanations that originate from the Absolute. This recognition is not an intellectual understanding but a direct, experiential realization that permeates every aspect of being.

At this level, the soul is no longer bound by the constraints of time and space. The distinctions between past, present, and future fade, and the individual experiences the eternal presence of the divine in every moment. This state of being is often described as the timeless now, where all things are seen as interconnected and existing in perfect harmony with the divine will. The enlightened soul perceives the universe as a single, unified whole, with no separation between the individual self and the greater cosmic reality.

The experience of enlightenment also brings with it a profound sense of inner peace—a peace that is unshakable because it is rooted in the eternal nature of the divine. The turbulence of the material world, with its cycles of birth, death, and change, no longer disturbs the soul, for it has come to rest in the unchanging truth of the Absolute. This peace is not a passive state but an active expression of divine love, radiating outward to all beings. The enlightened soul, having realized its oneness with the Source, becomes a channel through which the light of the higher emanations flows into the world.

Many spiritual traditions describe this state of enlightenment as the ultimate goal of the soul's journey. In the teachings of the Buddha, it is referred to as nirvana—the extinguishing of the flames of desire, ignorance, and suffering. In

Hindu philosophy, it is called moksha, or liberation, where the soul is freed from the cycle of rebirth and returns to the eternal oneness of Brahman. In Christian mysticism, it is the experience of divine union, where the soul becomes one with God and shares in the divine nature. Each tradition speaks of this ultimate state in its own language, but the underlying truth remains the same: enlightenment is the realization of the soul's inherent divinity and its eternal connection to the Source.

The examples of enlightened beings throughout history offer us glimpses of what this state looks like in the context of human life. Figures such as the Buddha, Jesus, and Rumi lived their lives as embodiments of the higher emanations, radiating wisdom, compassion, and love to all those around them. These masters did not retreat from the world but engaged with it from a place of deep spiritual realization, showing that enlightenment is not an escape from reality but a transformation of it. Through their teachings and actions, they demonstrated that the highest spiritual attainment is not isolation from the material world but the integration of divine awareness into every moment of existence.

For these enlightened beings, the flow of emanations is not something that happens outside of them but through them. They act as conduits for the divine energy that flows from the Source, bringing light, healing, and wisdom into the world. Their presence alone has the power to transform those around them, for they are living embodiments of the higher emanations. This is the essence of enlightenment: to become so aligned with the divine that one's very existence reflects the light of the Source.

The enlightened state is also marked by an expanded sense of compassion. The individual no longer sees others as separate beings but as expressions of the same divine essence. This recognition naturally gives rise to a deep love for all living beings, for the enlightened soul understands that the suffering and struggles of others are not external to them but part of the same journey of return. Compassion, in this sense, is not an emotion but a way of being—an expression of the unity that underlies all creation.

The enlightened soul moves through the world with a sense of detachment, not in the sense of indifference, but in the sense of freedom from the limitations of material existence. They are no longer bound by the desires and fears that once governed their actions. Instead, their actions flow from a place of divine inspiration, guided by the higher emanations. This detachment allows them to act with clarity and purpose, unburdened by the illusions of the material world. They are fully present in the world, yet their awareness remains rooted in the eternal reality of the Source.

In this final stage of enlightenment, the soul not only returns to the Source but participates in the ongoing process of emanation and return. The enlightened being becomes an active participant in the cosmic cycle, helping to guide others along the path of awakening. This is the role of the spiritual teacher or guide—a soul who, having reached the heights of enlightenment, turns back to help others on their journey. These beings are often described as bodhisattvas in the Buddhist tradition, those who delay their own final liberation in order to assist all beings in achieving enlightenment.

The enlightened state is also a state of creative expression. As the soul aligns with the flow of higher emanations, it becomes a channel through which the divine can manifest in the world. This creative expression can take many forms—art, music, poetry, teaching—but in every case, it is an outpouring of the divine energy that flows through the enlightened soul. The individual no longer creates from a place of ego or personal desire but from a place of divine inspiration. Their creations reflect the harmony and beauty of the higher realms, offering those who encounter them a glimpse of the divine.

In the Theory of Emanations, enlightenment is not the end of the soul's journey but a new beginning. The soul, having returned to the Source, now participates in the ongoing process of creation and return, helping to guide the flow of emanations and bring the light of the divine into the material world. This participation is both a responsibility and a gift, for the enlightened soul understands that its role is not to escape the world but to transform it. Through their actions, words, and presence, these enlightened beings help to raise the consciousness of all those around them, guiding them back to their own divine nature.

Ultimately, the process of enlightenment is a return to the truth of who we are: divine emanations of the Absolute, reflections of the Source in the world of form. The journey through the emanations, from descent into the material world to the ascent back to the Source, is the soul's way of rediscovering its true nature. Enlightenment is the full realization of this truth— a state of being where the soul lives in harmony with the flow of divine energy, participating in the ongoing creation and return of the universe.

In conclusion, the enlightened being lives in a state of union with the Source, reflecting the light of the higher emanations in every thought, word, and action. They are a living expression of the divine, fully aware of their place within the cosmic order and dedicated to helping others realize their own divine potential. This state of enlightenment is the fulfillment of the soul's journey, a return to the unity that underlies all things, and a participation in the eternal cycle of emanation and return.

Epilogue

As you conclude this journey through the emanations, you now find yourself in a different place from where you began. Something within you has changed, even if subtly. Perhaps it was not an abrupt revelation, but a gradual awakening, a recognition of something ancient and profound that was always there, waiting to be perceived.

After all, what have we learned about the universe, if not that it is both a mirror and a path? A mirror that reflects our own inner state, and a path that invites us back to the point of origin, to the absolute. But the crucial point is not in reaching that final destination. It is in the journey itself, in the process of returning to unity, in the act of remembering that we are part of something much greater and more vast.

The light that permeates everything, the invisible thread that connects every emanation, is also present within you. What was external and incomprehensible now seems closer, perhaps even within your grasp. From this point on, your life, just like the universe, can be seen through a new lens—that there is no real separation between the divine and the mundane, between the spiritual and the material. Every aspect of reality carries within it a portion of the original essence, veiled, yes, but always present.

The great mystery is not solved, but its purpose becomes clearer. The cycle of creation and return, of descent and ascension, continues. You, now aware of your role, are an active part of this cycle, and the choices you make, the intentions you hold, influence the flow of this cosmic energy. What will you do with this new knowledge? How will it shape your perception of the world and of yourself?

Remember, every step toward unity is significant, even if small. The journey does not need to be taken in grand leaps; it can be subtle, internal, like a stream that quietly flows between rocks, shaping the landscape over time. The path of return is not only about transcendence but about transformation—of yourself, of the world around you, and of how you choose to interact with the vast fabric of reality.

The deepest truth that this book has revealed is not something that can simply be explained in words, but something that must be felt, experienced. The universe is vast, yes, but every part of it is connected by the same flow of emanations that began at the origin of everything and continues to pulse in every atom, in every soul.

Now, as you close these pages, the process does not end. It only transforms. The ideas you have absorbed begin to take root, to blend with your own experiences and perceptions. And suddenly, you realize: what you read here is not just a theory or philosophy. It is a living truth, a truth that was already within you, waiting to be activated.

The cycle continues, the return is inevitable, and the great cosmic dance of creation and return does not cease. But now, you see it with new eyes. You feel it more deeply. And that, in itself, is the beginning of a new journey. A cycle within a cycle. A new stage of your own evolution.

Milton Keynes UK
Ingram Content Group UK Ltd.
UKHW042245011124
450424UK00001BA/249